Arsenical Pesticides

E. A. Woolson, *Editor*

A symposium sponsored by
the Division of Pesticide
Chemistry at the 168th
Meeting of the American
Chemical Society,
Atlantic City, N. J.,
Sept. 9, 1974.

ACS SYMPOSIUM SERIES **7**

AMERICAN CHEMICAL SOCIETY

WASHINGTON, D. C. 1975

Library of Congress CIP Data

Arsenical pesticides.
(ACS symposium series; 7)

Includes bibliographical references and index.

1. Pesticides—Environmental aspects—Congresses. 2. Arsenic compounds—Environmental aspects—Congresses.
I. Woolson, E. A., 1941– ed. II. American Chemical Society. III. American Chemical Society. Division of Pesticide Chemistry. IV. Series: American Chemical Society. ACS symposium series; 7. [DNLM: 1. Arsenicals—Congresses. 2. Pesticides—Congresses. QV294 A512 1974]

QH545.P4A77 574.5′222 74-31378
ISBN 0-8412-0243-5 ACSMC8 7 1-176 (1975)

PRINTED IN THE UNITED STATES OF AMERICA

ACS Symposium Series

Robert F. Gould, *Series Editor*

FOREWORD

The ACS SYMPOSIUM SERIES was founded in 1974 to provide a medium for publishing symposia quickly in book form. The format of the SERIES parallels that of its predecessor, ADVANCES IN CHEMISTRY SERIES, except that in order to save time the papers are not typeset but are reproduced as they are submitted by the authors in camera-ready form. As a further means of saving time, the papers are not edited or reviewed except by the symposium chairman, who becomes editor of the book. Papers published in the ACS SYMPOSIUM SERIES are original contributions not published elsewhere in whole or major part and include reports of research as well as reviews since symposia may embrace both types of presentation.

CONTENTS

PREFACE

Arsenic is an environmentally ubiquitous Group Va element. Although it has a metallic allotropic form like the two higher Group Va elements, antimony and bismuth, arsenic's chemical behavior is quite similar to phosphorus. It has four valence states, −3, 0, +3, and +5. Arsine and methylarsines are characteristic of As in the −3 valence state and are generally unstable in air. Arsenic metal is formed by reduction of arsenic oxides and is a glossy black material. Arsenic trioxide is a product of smelting operations, and it is the starting material for synthesizing most arsenical compounds. It is oxidized catalytically or by bacteria to arsenic pentoxide or orthoarsenic acid (H_3AsO_4). Arsines and arsenites are generally more toxic than arsenates.

In the natural environment, most arsenicals degrade or weather to form arsenate, although under anaerobic conditions arsenite may form. Biotransformations may occur which result in volatile arsenicals. These, in turn, are returned to the land where soil adsorption, plant uptake, erosion, leaching, reduction to arsines, or other processes occur. This natural arsenic cycle reflects a constant shifting of arsenic into the various environmental compartments.

This symposium was arranged because of current concern and interest in the quality of our environment, particularly as it pertains to trace heavy metals. While arsenic is not a heavy metal, it is a metalloid which has always generated much interest and controversy because many of its compounds have toxic properties. This symposium, then, was intended to examine the analysis, soil behavior, biotransformations, accumulations, cycling, and modeling of arsenate and arsenical pesticides in nature.

E. A. Woolson

Beltsville, Md.
November 1, 1974

1

Review of Arsenical Pesticides

S. A. PEOPLES

School of Veterinary Medicine, University of California, Davis, Calif. 95616

It is not the purpose of this paper to discuss the comparative usefulness of various arsenic compounds as herbicides and insecticides. It is rather to present information which is needed to form the basis for evaluating the effects of such usage on the ecosphere and the possibility of adverse effects on animals and man. Arsenic has had such a bad name as to be nearly synonymous with the word "poison" and I think it would be worth putting it in its proper place. It is a relatively common element, present in air, water, soil, plants and animals, and the pharmacology of chemical compounds depends on the dose given; no action, useful and toxic.

Considering that arsenic compounds have been known since 2500 B.C. and used in medicine since the time of Hippocrates since 400 B.C. (1), one would expect that information would be available as to the chemical nature of arsenic compounds in nature and their biochemical effects and metabolic fate in plants and animals. Such is not the case, however, due in large part to the failure to develop analytical procedures for specific compounds.

All samples were wet or dry ashed to inorganic arsenic, reduced to arsine with zinc and hydrochloric acid which was then quantitated by some modification of the Gutzeit method. The results were usually expressed as arsenic trioxide, a practice which has lead many to the conclusion that the arsenic in the sample was indeed trivalent arsenic and worse yet, toxicologists frequently use the term "toxicity of arsenic for animals" as synonymous with the toxicity of arsenic trioxide for animals (2).

Recently, an analytical method has been developed in my laboratory which will separately quantitate the inorganic and methylated arsenic in water, grass and urine (3) and is now being modified for soil samples. A method using a helium plasma has been introduced by Braman which will separate trivalent and pentavalent inorganic arsenic and methyl and dimethyl arsonic acids (4). The use of these methods in combination with such isolation technics such as thin layer and column chromatography will hopefully yield as much information about the biochemistry of arsenic as is now known about phosphorus.

Chemistry

Arsenic occurs in nature as sulfides such as orpiment and as complex sulfides or iron, nickel and cobalt. It has valences of 3^+, 5^+ and inorganic and organic compounds of both are known. A valence of -3 is also known where arsenic is combined with hydrogen in compounds known as the arsines.

Since the synthesis of salvarsan by Ehrlich in 1905, literally hundreds of organic arsenicals have been synthesized for use in medicine and industry. Except for the treatment of such parasitic diseases such as trypanosomiasis, amebiasis and filariasis, the advent of the antibiotics rendered most of the medicinal products obsolete but it is interesting to consider some of them since they have found new uses under new names, as pesticides and feed additive.

The information in Table 1, taken from pharmacology textbooks used in the 1930's shows some of these compounds and the therapeutic dose levels should be of particular interest to those concerned with their residue levels in food (5) (6). For example, it would be necessary to eat 10 kg of food containing 5 ppm of cacodylic acid to obtain a therapeutic dose.

Distribution of Arsenic in Nature

Arsenic is a ubiquitous element present in air, water, soil and all living tissues. The finding of arsenic in any sample is therefore not significant unless compared to the concentration normally expected. The range of these values is given in Table 2 (7). It must be pointed out that these values are given as total arsenic and their chemical nature is unknown. That high values occur in sea food has been known since 1935 when Coulson reported it in shrimp and found it to be nontoxic to rats (18). It is clear that animals and man receive a daily intake of arsenic which varies with geographical location and type of diet. Once again it should be noted that the chemical nature of the arsenic in these sources is largely unknown and therefore the fate and effect in animals and man cannot be predicted. For example Lakso (3) found that most of the background arsenic in Johnson grass was a methylated arsonic acid, probably MSMA, a compound of low toxicity.

Absorption, Distribution and Excretion

Arsenic compounds can be absorbed by any route although the usual entry is by ingestion. There is a common belief that arsenic is a cumulative poison which is largely based on its tendency to accumulate in high concentrations in the hair and nails, which are actually excretory products and not in equilibrium with the living body. A second basis for this belief is the

TABLE 1

Arsenical Drugs Used in Treating Human Disease

Drug	Dose grams	Use	Formula
Fowler's Solution	.005	Leukemia Tonic	K O As = O
Pearson's Solution	.005	Tonic	Na O As(=O)(CH_3)(CH_3)
Arrhenal	.05	Tonic	Na OAs(=O)(CH_3)(CH_3)
Sodium Cacodylate	.05	Tonic	Na O As(=O)(CH_3)(CH_3)
Arsphenamine	.3-.6	Syphilis	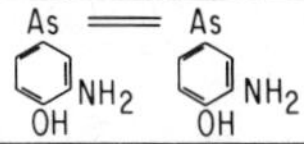
Atoxyl	.02-.20	Trypanosomiasis	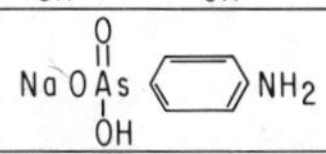
Tryparsamide	2.0	Trypanosomiasis	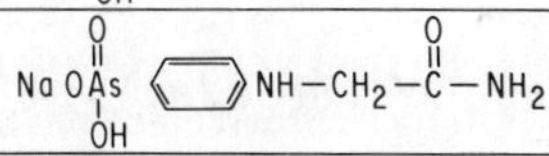
Carbarsone	.75	Amebiasis	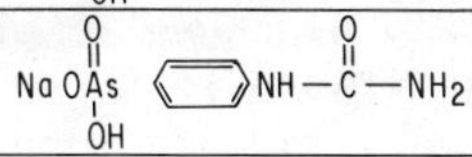
Melarsoprol	.18	Trypanosomiasis	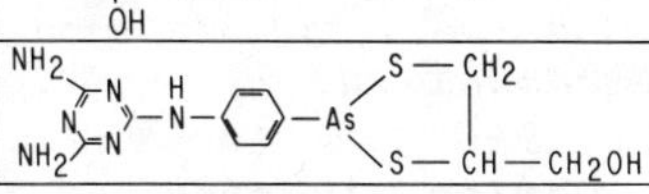

Pharmacotherapeutics (5)
A Manual of Pharmacology (6)

Table II

Concentration of Arsenic in Nature

Substance	Concentration ppm
Water	.01 - 1.0
Soil	1.0 - 500.0
Grass	.1 - 1.6
Ferns	.2 - 3.64
Vegetables	.00 - 2.9
Grains	.11 - .16
Corn oil	.00 -
Fish oil (ocean)	1.0 - 5.0
Fish	2.0 - 9.0
Shellfish	1.6 - 2.9
Shrimp, lobster	1.5 - 100.0
Meat	.06 - 1.07
Milk	.01 - .05

J. Chronic Diseases (7)

use of the rat as an experimental animal in arsenic studies, an animal which is unique in its ability to store arsenic in its red blood cells.

Table 3 (9) shows the normal values in the tissues of common experimental animals and their response to the feeding of arsenic. If man were included in the table he would have values similar to those of the rabbit with values of all tissues varying between .05 and .30 ppm. Milk is exceptionally low in all animals, being .01 - .03 ppm.

Hair and nails normally have 1.0 - 5.0 ppm but due to the fact that they readily pick up arsenic from the environment, such values are of uncertain value in reflecting poisoning. Blood values are useful in measuring the body burden and rate of change in the body as shown in Fig. 1 (10). The clearance rates clearly illustrate the differences between animal species and the extremely slow rate in the rat.

The route of excretion is in the urine and feces and the shape of the excretion curve in the rate as shown in Fig. 2 (11), showing that there is a fast phase followed by a slow one, indicating that there are two storage pools.

The arsenic level in the milk of cows does not increase with the blood concentration when fed inorganic arsenic, methyl arsonic acid or cacodylic acid as illustrated in Fig. 3 (12). This may mean that there is an active transfer of arsenic in the mammary gland which is saturated at all times. From a residue point of view, the presence of even toxic amounts of arsenic in animal feed will not effect the safety of milk. The urinary excretion of arsenic in urine has been suggested as a measure of exposure but it is of no value unless the diet is known. Schrenk (13) found that the eating of sea food, particularly crustacea, raised the urinary arsenic levels by as much as 10-fold, returning to normal in 20 - 40 hours as shown in Fig. 4. This is due to the compound known as Coulsons "Shrimp Asenic" and illustrates its rapid absorption and excretion.

Metabolism of Arsenic

The preceeding studies on the distribution and excretion of arsenic are based on total arsenic determinations and say nothing of the chemical changes that might occur. Gosio (14) noted in 1893 that certain molds growing on wallpaper containing arsenical pigments produced a toxic gas which was identified by Challenger (15) in 1931 as trimethyl arsine and could also be produced from sodium arsenite and cacodylic acid. Recently McBride (16) reported that a methanobacterium could methylate inorganic asenic to form dimethyl arsonic acid and dimethyl arsine under anaeorbic conditions. This finding suggested to us that the methyl arsonic acid we had found in bovine urine (17) need not entirely come from grass which contains methylated arsenic (3) but could have been produced by the anaerobic bacteria in the rumen. Preliminary

TABLE 3

THE ARSENIC CONCENTRATION IN TISSUES AFTER 21 DAYS OF FEEDING A DIET CONTAINING 50 PPM ARSENIC TRIOXIDE

Animal*		Tissue concentration of arsenic ppm									
		Liver	Heart	Kidney	Spleen	Fat	Muscle	Brain	G.I. tract	Skin	Blood
Rat	Control	0.0	3.3	1.5	0.67	0.6	0.7	0.5	0.6	0.6	15.0
	Fed	20.0	43.0	25.0	60.0	12.0	3.0	3.8	15.0	27.0	125.0
Guinea Pig	Control	0.0	0.0	0.0	0.0	0.0	0.0	0.0	1.0	0.0	0.0
	Fed	1.0	20.0	1.0	15.0	0.8	2.0	0.0	2.0	—	4.0
Rabbit	Control	0.0	0.0	0.0	0.0	0.0	0.0	0.0	0.1	0.0	0.1
	Fed	1.0	0.2	1.5	0.2	0.2	0.2	0.0	1.5	2.5	1.5
Hamster	Control	0.0	0.0	0.0	0.0	0.0	1.8	0.0	1.5	0.0	0.0
	Fed	15.0	7.0	5.0	2.0	0.7	2.5	1.0	30.0	38.0	2.5

* Four in each group.

Ann. N. Y. Acad. Science (9)

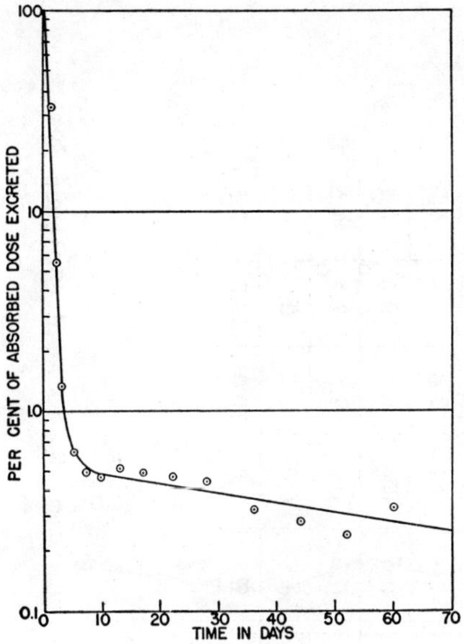

Proceedings of the Society for Biology and Medicine

Figure 1. ^{76}*Arsenic levels in whole blood. Concentration of* ^{76}*arsenic per gram of blood at a particular time is expressed as per cent of the administered dose per gram of body weight* (10).

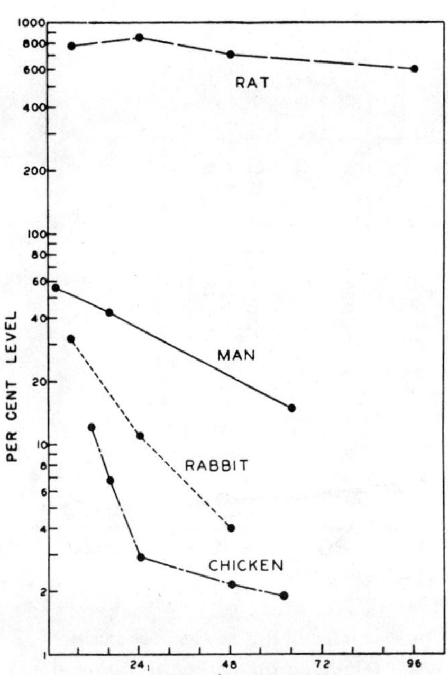

University of California Publications in Pharmacology

Figure 2. Total fecal and urinary elimination of carrier-free ^{74}*As by rats* (11)

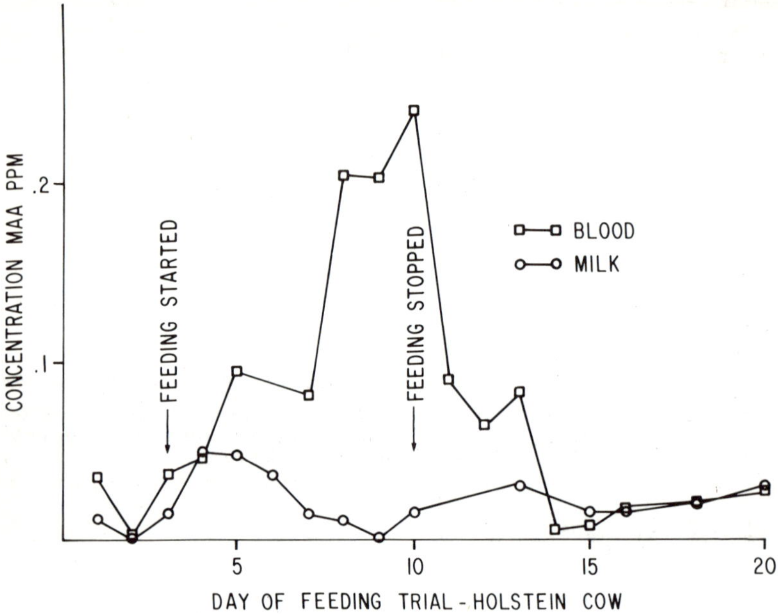

Figure 3.

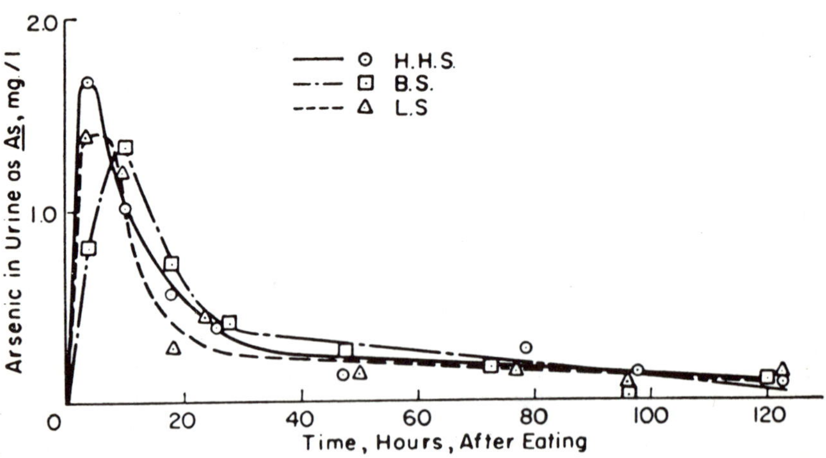

American Industrial Hygiene Association Journal

Figure 4. Concentration of arsenic in urine following the eating of lobster

experiments have shown that the feeding of either sodium arsinate or sodium arsinite results in a great increase in methyl arsonic acids in the urine, a finding which seemed to confirm this hypothesis. To show that the rumen was necessary for this methylation the same experiments have been repeated in dogs, and if confirmed by further experiments, the carnivore can methylate inorganic arsenic as well as the ruminant.

Experiments are now under way to find the role, if any, of the rumen, and the biochemical mechanism of the methylation of arsenic in the dog. The origin of methyl arsonic acid in normal plants is still to be explored, but one could postulate that it could be of animal origin.

The work of Braman (4) indicated that methylated arsenic compounds are probably quite common in man and nature and only a small beginning has been made in understanding the arsenic cycle.

It should be pointed out that unlike mercury, methylation causes a great reduction in toxicity and is a true detoxication process.

Toxic Effects of Arsenic

The toxicity of trivalent inorganic arsenic for living organisms varies widely as shown in Table 4.

Table 4

Item	Arsenite	Arsenate
Bacteria	290	> 10,000
Algae		> 1,000
Yeast		300
Daphnia magna	5.2	12.5
Flatworms	40	361
Minnows	20	250
Minnows	17.8	234
Minnows	11.6	60
Rats and mice, 96 hr LD_{50} oral	11.2	112
MLD, intraperitoneal	5.8	21

J. Chronic Diseases (7)

The toxic dose for rats of arsenical herbicides is shown in Table 5, which are average values from a variety of sources.

The mechanism of toxicity is considered to be due to the binding of - SH groups of lipoic acid by trivalent arsenic. However, pentavalent arsenic compounds behave like phosphate in biochemical reactions yet symptoms of poisoning are similar for all arsenic compounds. It may be found that they can be reduced to trivalent arsenic by mechanisms presently unknown. The fact

Table 5
Toxicity of Arsenical Herbicides for Rats

Compounds	Toxic Dose mg/kg
Sodium Arsenite	20 - 60
Sodium Arsenate	80 - 120
Sodium Methyl Arsenate (MSMA)	1200 - 1600
Sodium Cacodylate	1200 - 1600

that BAL, 2,3 dimercaptopropanol, is used with some success in all forms of arsenic poisoning is further evidence for this view. Acute toxicity is the usual form of poisoning, with the development of acute enteritis and death in shock. With lower doses, there is extensive liver and kidney damage as well. The possibility of chronic arsenic poisoning from continuous ingestion of small doses is rare, due to the detoxication and excretion being fairly rapid.

Arsine is a gas formed in the presence of acid and hydrogen and is toxic in very low cencentrations in inhaled air. It causes hemolysis of the red cells, resulting in anemia, hemoglobinuria with resulting renal damage. It is rapidly fatal and there is no known antidote. As previously noted, McBride (16) has shown that dimethyl arsine can be produced in the bacteria found in the mud of waterways and suggested it could cause toxicity. This possibility seems remote since dimethyl arsine burns spontaneously in air.

The question of arsenic as a cause of cancer has been studied for a century without a clear decision. When arsenic trioxide is applied to the skin or taken orally in large doses hyperkeratotic lesions appear on the skin with a delay of many years in some cases. These lesions are precancerous but they can also be caused by sunlight, x-ray and thermal burns. Cancer has not been produced in experimental animals by any arsenic compound. The incidence of cancer in arsenic industries is equivocal is spite of continuing studies (18).

Conclusion

The use of new analytical methods will hopefully elucidate the role of arsenic in the ecosphere and may even demonstrate that it is essential in living tissue. The high levels in sea animals which are constant in value and not related to industrial or agricultural pollution, suggest this possibility. While much research needs to be done, enough is known about the useful and toxic effects of arsenic compounds so that their use as pesticides and feed additives can be carried out with safety.

Literature Cited

1. Vallee, B.L., Ulmer, D.D., and Wacker, W.E.C. "Arch. of Industrial Health." Arsenic Toxicology and Biochemistry 21:132-151, (1960).

2. Luh, M.D., Baker, R.A., and Henley, D.E. "Arsenic Analysis and Toxicity - A Review." Science of the Total Environ. 2:1-12, (1973).

3. Lakso, J.U., Peoples, S.A., and Bayer, D.E. "The Simultaneous Determinations of MSMA and Arsenic Acid in Plants." Weed Science 21:166-169, (1973).

4. Braham, R.S., and Forebach, D.D. "Methylated forms of Arsenic in the Environment." Science 182:1247-1249, (1973).

5. S. Soles-Cohen and T.S. Githens, Pharmacotherapeutics, D. Appleton, New York, (1928).

6. T. Sollman, A manual of Pharmacology, 7th Ed., W.B. Saunders, Philadelphia, (1948).

7. Schroeder, H.A., and Balassa, J.B. "Abnormal Trace Elements in Man:Arsenic." J. Chronic Diseases 19:85-106, (1966).

8. Coulson, E.V., Remington, R.E., and Lynch, K.M. "Metabolism in the Rat of the Naturally Occurring Arsenic of Shrimp as Compared with Arsenic Trioxide." J. Nutr. 10:255-270, (1935).

9. Peoples, S.A. "Arsenic Toxicity in Cattle." Ann. N. Y. Acad. Science 111:644-649, (1964).

10. Ducoff, H.S., Neal, W.B., Straub, R.L., Jackson, L.O., and Brues, A.M. "Biological Studies with Arsenic II Excretion and Tissue Localization." Proc. Soc. Exper. Biol. Med. 69:548-554, (1948).

11. Lanz, H., Jr., Wallace, P.W., and Hamilton G. "The Metabolism of Arsenic in Laboratory Animals Using 74 As as a Tracer. Univ. of Calif. Pub. in Pharmacol. 2:263-282, (1950).

12. Peoples, S.A., Unpublished.

13. Schrenk, H.H., and Schreibeis, L., Jr. "Urinary Arsenic Levels as an Index of Industrial Exposure." Am. Ind. Hyg. Assoc. J. 19:225-228, (1958).

14. Gosio, B. Arch Ital. Biol. 18:253-298, (1893).

15. Challenger, F. Biological Methylation. Chem. Rev. 36: 315-361, (1945).

16. McBride, R.C., and Wolfe, R.S. "Biosynthesis of Dimethyl Arsine by Methanobacterium." Biochem. 10:4313-4317, (1971).

17. Peoples, S.A., Lakso, J., and Lais, T. "The Simultaneous Determination of Methylarsonic Acid and Inorganic Arsenic in the Urine." Proc. West. Pharmacol. Soc. 14:178-182, (1971).

18. E. Browning, Toxicity of Industrial Metals, Appleton-Century-Crofts, New York, (1969).

2

The Determination of Traces of Arsenic: A Review*

YAIR TALMI and CYRUS FELDMAN

Oak Ridge National Laboratory, Oak Ridge, Tenn. 37830**

This review is organized in terms of the main steps and topics characterizing all trace analysis: sample pretreatment and dissolution, solution stability, preconcentration, isolation and determination. The options now existing in each step have been presented, but no attempt is made to mention every investigation in which each has been used. It is hoped that this treatment will make it easier for the analyst to assemble a procedure suited to the needs of a particular case.

Pretreatment and Dissolution of the Sample

If total arsenic is to be determined, the sample must be mineralized at least to the degree necessary to convert all arsenic present to inorganic forms. The treatment chosen should be the mildest treatment which will accomplish this conversion, so that losses and contamination will be minimized. Several approaches are possible:***

1. Wet Ashing. According to Portmann and Riley[1], prolonged digestion with nitric acid, followed by evaporation to dryness (hot-plate surface temperature ∿180°) will give no losses of arsenic, even if several mg of chloride were originally present. This type of attack could be used for materials such as vegetation

---

*Research supported by the National Science Foundation--RANN (Environmental Aspects of Trace Contaminants) Program under NSF interagency agreement No.389 with the U.S. Atomic Energy Commission.

**Operated for the U.S. Atomic Energy Commission by Union Carbide Corporation under Contract No. W-7405-Cong.-26.

***A general and more detailed treatment of this subject is given in T.T. Gorsuch's book "The Destruction of Organic Matter", Pergamon, NYC (1970).

and animal muscle. Stronger treatment is usually advocated for other organic materials, especially those high in lipids; however, Kingsley and Schaffert[2] state that all of the arsenic present in liver, kidney or muscle can be recovered merely by digesting the blenderized tissue for 30 minutes in ∿4N HCl. The digestion must be open to the atmosphere, so that sulfides and mercaptans (which may interfere with subsequent determination of As) will be eliminated. Similarly, R.F. Abernethy and F.H. Gibson[3] state that essentially 100% of the arsenical species in coal can be extracted by boiling the powdered coal gently in a 1:7 mixture of HNO_3 and H_2O. The HNO_3 may then be eliminated, if desired, by adding 1:1 H_2SO_4 and evaporating the solution to fumes. According to R.B. Baird, S. Pourian and S.M. Gabrielian[4], raw sewage and primary and secondary effluents can be mineralized by refluxing with 4% HNO_3 + 1% H_2O_2, followed by evaporation and further additions of HNO_3. Cacodylates sprayed on sandy soil can be recovered by 5 minutes of shaking with a mixture of strong H_2SO_4 and HCl plus decolorizing carbon. Hamme, Young and Hunter[5] state that recovery by this rapid method averages 93%.

Other treatments have also been advocated which do not necessarily mineralize biological samples completely, but liberate their metals to a degree sufficient for some types of analysis. W. J. Adrian[6], following G.W. Gordon, left the sample overnight, with HNO_3 and $HClO_4$, in a tightly sealed polyethylene bottle. The bottle was then placed in hot running water for 2-3 hours, cooled and opened. Rapid dissolution of tissues was also achieved by A. Bouchard[7] using concd. H_2SO_4, CrO_3 and red fuming HNO_3 (possible contamination from reagents must be considered in this case). Many tissues can also be transformed into clear solutions with the aid of tetramethylammonium hydroxide, either in solid or aqueous solution form[8] or as a solution in toluene[9]. Heating at 60°C may be required. Many of these milder procedures do not attack fatty and high-lipid tissues or bone and tooth specimens.

When stronger treatment is needed, some combination of nitric, sulfuric and perchloric acids is usually used. In a typical procedure of this type, Chu, Barron, and Baumgarner[10] alternately heat the sample in a mixture of sulfuric and nitric acids until the solution darkens, then add more HNO_3, and repeat the process until darkening no longer occurs; they complete the oxidation by heating with $HClO_4$. Sandell[11] recommends the use of refluxing with such procedures in order to prevent the loss of As_2O_3 and/or $AsCl_3$. If the arsenic is eventually to be reduced to As^{+3} for determination, the excess HNO_3 remaining after wet-ashing can be destroyed by treating the mixture with a few ml of saturated ammonium oxalate and warming. G.I. Spielholtz, G.C. Toralballa, and R.J. Steinberg[12] rapidly (and safely) mineralized powdered coal by refluxing it with a mixture of 68-70% $HClO_4$ and para-HIO_4.

P. Schramel[13,14] showed that reagent blanks in the dissolution of 150 mg tissue samples can be minimized by warming the sample with 100 µl concd. H_2SO_4 and adding 50% H_2O_2 solution dropwise.

This is a refinement of W. Migault's "Caro's acid" procedure[15].

2. Dry Ashing. Various procedures involving $MgO/Mg(NO_3)_2$ addition have been recommended for dry-ashing vegetable and animal tissues. One recent, widely applicable procedure suggested by George, Frahn and McDonald[16] involves preliminary blenderizing of the tissue, treatment with MgO and cellulose powder and an initial cautious charring in a porcelain crucible. When cool, the crucible is treated with $Mg(NO_3)_2 \cdot 6H_2O$ and placed in a cold muffle furnace. The temperature of the furnace is slowly raised to 555°, and kept there for 2 hours. Good recovery was obtained for 1-2 ppm spikes added to samples. R.F. Abernethy and F.H. Gibson[3] obtained quantitative recovery of As from coal by igniting it with MgO at 650°C; the residue was dissolved in 7N H_2SO_4.

3. Oxygen Combustion. The Schöniger combustion method[17,18] (ignition of a sample in a closed flask containing O_2 and an absorbing solution) was used on dried tissue by Schwedt and Russel[19]. They obtained quantitative recovery of As on 0.5 g dried specimens with a 750 ml flask containing O_2 and 5 ml of 3N HCl. Ignition in a metal bomb containing O_2 was also found satisfactory by H.S. Saterlee and G. Blodgett[20]. According to C.E. Gleit and W.D. Holland[21], biological tissues, as well as other substances can be mineralized at a low temperature by using electrically excited oxygen. Complete recovery of As is obtained from blood treated with $HAsO_2$.

4. Fusion. Minerals are usually fused with NaOH in a silver or nickel crucible; e.g., by H. Onishi and E.B. Sandell[22]. These authors state that losses of As are 0.5%; essentially all of the As is recovered in the leach liquid, even when a residue is present. J.A. James and D.H. Richards[23] have also applied this fusion to the determination of As in elemental Si.

Stability of Sample Solutions During Storage

This discussion pertains mainly to natural water samples and standards, since the sample solutions produced by any of the above methods are normally analyzed soon after preparation.

To test the stability of As at low concentrations in sea water, J.E. Portmann and J.P. Riley[1] filtered each sample through a 0.5μ Millipore filter, and treated the filtrate with 5 Ci of carrier-free ^{74}As. Samples in soda-glass bottles levelled off at 16% loss after 16 days; samples in polyethylene and borosilicate glass levelled off at 6% loss after 10 days. These authors advise storage of such samples in frozen form in polyethylene containers. G.C. Whitnack and R.G. Brophy[24] made small additions of Na_3AsO_3 solution to well-water samples, and stored them in 25 ml polystyrene vials with polyethylene caps. No loss of As(III) from these solutions was detectable after one week. A.S. Al-Sibbai and A.G. Fogg[25] measured the stability of solutions of both As(III) and As (V) in various containers at the 4-20 μg/ml level. Samples of As_2O_3 were dissolved in NaOH solution and the solution neutralized. This solution kept its full titer for 56 days in borosilicate glass, soda glass and polyethylene containers, in both light and dark storage areas. Solutions of $NaHAsO_4 \cdot 7H_2O$ in water did the same for 100 days under the same conditions. R.S. Braman[26] notes that very low concentrations of As species tend to disappear rapidly from natural water samples. In our laboratory[27], however, no losses of $^{74}As(V)$ were experienced in distilled water (polyethylene and soft glass containers) 15% HNO_3 or 5% $HClO_4$ solution (soft glass containers) over a period of 3 weeks.

Preconcentration and Isolation of As Species

The methods described in this section are oriented toward the eventual determination of total (inorganic) As, regardless of the original molecular form in which the As occurs. The principal methods which have been used to preconcentrate As species are coprecipitation and adsorption, volatilization and liquid-liquid extraction.

1. Coprecipitation and Adsorption. $Fe(OH)_3$ has been known for some time to be an efficient collector of arsenate ion. For example, R. Pieruccini[28] collected 100 μg of As(V) from 24 liters of water (= 4 ng/ml) by using 150 mg of $Fe(OH)_3$ (the As was then determined spectrographically). J.E. Portmann and J.P. Riley[1], precipitating $Fe(OH)_3$ at pH 7, recovered 99% of the 2 μg of As(V) present in a liter of water (2 ng/ml). This procedure has also been used to collect As for determination by the Gutzeit[29] and xanthate[30] extraction procedures. V.I. Plotnikov and L.P. Usatova[31] carried out numerous coprecipitation experiments in 50 ml of solution. They found that at pH 7, As(V) is quantitatively carried down by the hydroxides of Ce, Zr, In, Fe, Ti and Al. The yield decreases, especially for Al at pH >8. In all cases, coprecipitation was more efficient than the addition to the As(V) solution of a prepared hydroxide slurry. For As(III), efficient (≥95%) coprecipitation was obtained only with In and Zr hydroxides at pH 8.5. W. Reichel and B. G. Bleakley[32] obtained complete recovery of 0.2-3.0 mg of As(V) (as well as similar amounts of Se, Te, Sb, Sn, Bi, Pb and Fe) from 20g of Cu by coprecipitating with $La(OH)_3$ at pH 9-10. P.M. Santoliquido[33] removed carrier-free $^{76}As(V)$ quantitatively from a 7N HNO_3 solution by passing it through a 7 x 40 mm column of hydrated MnO_2. The capacity of this column for As(V) is greater than 272 μg As(V)/g MnO_2. Z.G. Hanna[34] used $Mg(OH)_2$ as a coprecipitant: 0.3 μg As(V) in 10 ml was treated with $MgCl_2 + NH_4Cl$, and then with NH_4OH. The precipitate obtained was dried, and ignited at 600°C. No As was lost.

Thionalide is soluble in acetone, but insoluble in water. J.E. Portmann and J.P. Riley[1], making use of this fact, added an acetone solution of thionalide to a 0.5N H_2SO_4 solution of 50 ng As(III)+ ^{74}As tracer in 1 liter of sea water. The solution was stirred, boiled 30 min. to remove acetone, and allowed to stand overnight. The precipitate, consisting almost entirely of thionalide itself, was wet-ashed with concentrated HNO_3. Recovery of ^{74}As was 95%. Talmi, et al.[27] found that the high salt content of the sample solution seems to be necessary to insure complete precipitation;

however, overnight storage is unnecessary if the precipitation is performed at ∿0°C.

Small amounts of As(III) (reduced from As(V) if necessary with KI) were collected by a sulfide procedure by V.V. Sergeeva, I.S. Levin, L.I. Tishchenko and V.S. Dankova[35]. Interfering elements were first removed by cupferron precipitation and filtration. Thioacetamide was added to the filtrate, and hydrolized by heating the solution. The resulting As_2S_3 precipitate was collected by centrifugation; the yield was 85%.

2. Liquid-Liquid Extraction. A.K. Klein and F.A. Vorkes[36,37] were among the first to use xanthates to extract As(III) from food and biochemical tissues. In this case, the first collection was done with $Fe(OH)_3$; the final determination, with arsenomolybdate. P.F. Wyatt[38,39] used diethylammonium diethyldithiocarbamate in $CHCl_3$ to isolate As. Since this reagent does not extract As(V), several potentially interfering metals were removed initially by first performing this extraction with As in the pentavalent state. As was then reduced to As(III), and extracted from 1-10N H_2SO_4 solutions. Some potential interferences not eliminated in this way [Cu, Bi, Sb(III)] can be removed by a preliminary extraction with cupferron. According to T.J. Veleker[40], this reagent can be used to extract As(III) away from Ge, Sb and Bi in 6N HCl. H. Malissa and E. Schöffmann[41] found that As(III) could be precipitated by ammonium pyrrolidine dithiocarbamate (APDC), as well as by other dithiocarbamates at pH 2-6. C.E. Mulford[42] used a methyl isobutyl ketone solution of APDC to extract As(III) for determination by atomic absorption. V.V. Sergeeva, I.S. Levin, L.I. Tishchenko and V.S. Dankova[35], using an 0.5N solution of di-2-ethylhexyldithiophosphoric acid in decane and organic/aqueous volume ratios of 1:1 to 1:30, extracted traces of As(III) from 0.5N acid aqueous solutions. The As was recovered from the organic phase by shaking with bromine water. A.I. Busev and M.I. Ivaniutin[43] found that a similar reagent, diethyldithiophosphoric acid, used similarly, extracted As(III) from either weakly or strongly acidic aqueous solutions.

Perhaps the simplest of the procedures for segregating As is E. Gagliardi and H.P. Wöss's[44] extraction of $AsCl_3$ from 6-7N HCl into a mixture of 2 volumes of CCl_4 with 3 volumes of 2-butanone, 2-pentanone or 2-heptanone. As with other extractants, potentially interfering elements can be eliminated by conducting a preliminary extraction with the As in the pentavalent state. According to A.R. Byrne[45], AsI_3 can be extracted with toluene from a solution 12N in H_2SO_4 and 0.05 M in KI. It can be stripped from the toluene with 6N H_2SO_4 + 0.05 M KI.

3. Volatilization. The principal forms in which As is volatilized for analytical purposes are as a trihalide and as a simple or substituted trihydride (arsine). After biological material was decomposed with H_2SO_4 + HNO_3, E.B. Sandell[11] distilled $AsBr_5$, using a special still. Some P and Sn were also distilled. G.R. Kingsley and R.R. Schaffert[2] noted that arsenic compounds can be leached from tissue homogenates with 1+2 HCl, but that the operation must

be performed in the open, rather than under total reflux if arsenic is then to be distilled off as $AsCl_3$. This precaution is taken in order to expel H_2S and mercaptans which would otherwise tend to prevent the distillation of $AsCl_3$. Recent years have seen a rapid growth in the use of arsine evolution as a method of separating As from its original matrix. Most authors who have used Zn° in acid to generate AsH_3 (e.g. R.E. Madsen[46], F.J. Fernandez and D.C. Manning[47]) have first reduced the As to As(III) with $SnCl_2$ and/or KI. E.N. Pollock and S.J. West[48] used $TiCl_3$ for the preliminary reduction and Mg° in acid for generating AsH_3. F.E. Lichte and R.K. Skogerboe[49], however, inject a small volume of acid sample solution into a columm of granular Zn°; the sample is forced through the column by a stream of argon. The AsH_3 thus generated passes with the argon into the detection system. R.S. Braman, L.L. Justen and C.C. Foreback[50] used $NaBH_4$ to reduce As compounds to AsH_3. They find that they can discriminate between As(V) and As(III) by controlling the pH of the sample solution: As(V) is reduced only if pH ≤ 1.5. The $NaBH_4$ can be injected as a solution, or by dropping a pellet of the solid reagent into the sample solution by means of an externally operated hopper as suggested by F.J. Fernandez[51]. R.N. Sandell[11], as well as R.S. Braman, L.L. Justen and C.C. Foreback[50] caution that easily reduced ions such as Cu(II) may interfere with the production of AsH_3. After generation, the AsH_3 (b.p.-55°C) may either be passed directly into the detection device or accumulated in a liquid nitrogen-cooled trap and released to the detector over a very short period of time by warming the trap (see below).

Methods of Determination

1. Molecular Absorption--Spectrophotometry. There are a few reagents which produce intense-color derivatives with arsenic. However, two of them, silver-diethyldithiocarbamate (Ag-DDC) and ammonium molybdate are universally accepted as most suitable for spectrophotometric measurements. The Ag-DDC[52] reagent is usually used in conjunction with the arsine generation method[53,54]. Arsine is passed through 0.5% Ag-DDC solution in pyridine and the intensity of the red color is measured at 533 nm. Beer's law is obeyed over the 1-20 μg As range and the limit of detection is below 0.1 ppm. The arseno-molybdate complex is considered more suitable by many because of its sensitivity, reliability and general freedom from interferences[55]. Arseno-molybdic acid, formed by the reaction of arsenate with acidified molybdate, is reduced to the blue complex, the absorption of which is measured spectrophotometrically. $SnCl_2$ is used by a few workers for reduction[56-58], but produces an unstable color. Others use hydrazine sulfate[59,60] but reduction is slow. Portmann and Riley[1] found that a solution 0.4N in H_2SO_4 and 0.12% in ammonium molybdate will produce a very stable complex at room temperature within 30 minutes. Absorption is measured at 866 nm. No interferences were observed in the analysis of sea water, silicate rocks or marine organisms. The arsenic can be separated from the matrix via arsine generation[61,62], solvent ex-

traction[63,28,64,65] or coprecipitation with ferric hydroxide[28] or thionalide[1]. Stará and Starý[66] extract As(III) from sulfuric acid solutions following conversion to AsI_3. The compound develops an intense yellow color when treated with 8-mercaptoquinoline. Procedures were developed to prevent interferences from the oxidation of I^- to I_2 or from the formation of insoluble iodides. Mankova and Maksimenko[67] described a method based upon the reduction of $AgNO_3$ to Ag° by AsH_3. The Ag° exerts a catalytic effect on the further reduction (by Fe^{+2}) of $AgNO_3$. There is a relationship between the reaction rate, measured photometrically, and the concentration of the catalyst and thus the concentration of arsenic. The detection limit of the reaction is 20 pg/ml.

A few workers have developed methods for the separate determination of arsenite and arsenate species[63,68,62]. Spectrophotomeric methods have been applied to a large variety of samples including urine[69], blood and biological material[70], yeasts[71], soil[61], and atmospheric dust[72]. At the present time, spectrophotometry is still the most widespread technique for the determination of arsenic, mainly because of its inherent methodical and technical simplicity and its low cost.

2. Radiochemical Techniques.

A. Among the various radiochemical techniques, neutron activation analysis (NAA) is unique in its widespread applicability to the determination of arsenic. Although, in principle, NAA is a non-destructive analytical technique, radiochemical separation schemes are almost always required to avoid overlapping of various photo-peaks. Various such schemes[73-78] are based on a combination of two or more separation techniques, such as distillation, precipitation, solvent extraction or ion exchange. With the advent of high resolution solid state detectors, the direct instrumental NAA approach has been attempted. Unfortunately, the high activity of ^{24}Na induced in many environmentally based samples prevents the determination of arsenic at concentrations below a few ppm, since the γ-activity of these samples will have to decay for 4-5 days before measurement. The induced γ-activity of ^{76}As is measured by monitoring the 559 KeV photopeak. This peak will appear, in many samples, as the middle peak of a triplet composed of ^{82}Br ($t_{1/2}$ = 35.3h), ^{76}As ($t_{1/2}$ = 26.5h) and ^{122}Sb ($t_{1/2}$ = 67.2h) requiring the resolution performance obtained by Ge(Li) detectors. Also, since ^{76}As has the shortest half-life in the triplet, the counting should be done as soon as possible after irradiation. NAA is one of the most sensitive techniques with a detection limit of 0.1 ng using a thermal neutron flux of 10^{-12} neut.-cm^{-2}-sec^{-1}. The method is useful at the sub-ppm concentration level with sample sizes substantially smaller than those required by the colorimetric methods. Through the years, NAA has been successfully applied to a large variety of samples; biological and marine samples[78-87], plant tissues[88,89], water samples[90,91], air-borne particulate matter[92], pesticide distribution[93], heavy oil spills[94], geological samples[75,95,96], coal ash[31], and many others. Precision and accuracy for these samples are generally at the 3-10% range

B. Substoichiometric isotope-dilution. In 1964, a substoichiometric isotope dilution technique was reported[97,98] for the determination of traces of arsenic following activation of the sample. The method offers a simple, rapid and rather selective radiochemical separation of arsenic. Detection limits were at the 0.01[99] and 0.04[100] ppm level. The method was simplified to allow the determination of microgram amounts of arsenic without sample pre-irradiation[99]. The arsenic sample is wet-ashed, radio-labelled with ^{74}As standard solution, complexed with limited amounts of zinc-diethyldithiocarbamate and extracted by $CHCl_3$. Arsenic contents of organic and inorganic compounds were determined with average recovery better than 98.5%. However, this method cannot be easily applied to trace analysis of arsenic in complex matrices such as biological materials.

3. Electrochemical Techniques.

A. Amperometric titrations. Various potentiometric methods are available for the determination of arsenic with Ce^{+4}, MnO_4^-, IO_3^- [101,102] or with Dichloramine-T[103]. Arsenic in organic materials can also be determined by potentiometric titration[104,105]. Errors are typically 1% and concentration ranges of 0.020-100 mg are typical. The technique is not useful for trace levels of arsenic.

B. Polarography. Arnold and Johnson[106] have studied and reviewed the behavior of arsenic in various media and the mechanisms of the electrode reactions using DC and square wave polarography. They describe the determination of arsenic at the 5% level in iron ore. Copper was determined at the 0.02-1 mg level of arsenic with a rsd of ±3%. Arsenic in lead was determined at the 30-5000 ppm level, and in zinc at the 10-1000 ppm level with rsd of ±2%. In biological materials, the practical As level is 1-1000 μg. The detection limit of the technique is thus generally 1-10 ppm, but below 1 ppm accuracies and precisions are much poorer. Whitnack and Brophy[23] described a very sensitive and rapid method for the determination of arsenic in drinking water. The method utilizes a single-sweep polarographic technique with detection limit of 5 ppb and rsd of 5-10%. No special pretreatment of the sample is necessary. However, the requirement of an oscilloscopic readout or a sophisticated data acquisition system seems to have prevented the widespread use of the technique. In addition to DC and square wave polarography[107], arsenic was also determined by anodic stripping from a Pt electrode[108] and by differential pulse polarography, DPP[109]. More recently, Meyers and Osteryoung have further improved the application of DPP to arsenic determination[110]. The detection limit under optimized conditions is 0.3 μg/l and the response is linear to 500 mg/l. The rsd is 1.4% for 19.3 μg/l and 16.4% for 1.93 μg/l. A procedure (applied after the arsenic polarogram is obtained) was devised to correct for interferences from Pb, Sn^{IV}, Tl^{I}, or Tl^{IV} which was satisfactory down to 20 μg/l. No applications were described, although the method was claimed to have been applicable to biological samples, e.g. unfiltered activated sludge.

C. Coulometry. The determination of arsenic by coulometry was described by Bruckenstein and Johnson[111] and by Simon, Christian and Purdy[112] who have applied coulometric titration to arsenic determination in urine. Their method was later adopted by Woolson, Axley and Kearney[61] for the determination of arsenic in soil. It was found that various salts present in soil do not interfere with the analysis. Comparison between colorimetric and coulometric methods indicated that there is no apparent difference between their accuracy, although the coulometric method was more precise.

4. X-ray Fluorescence Spectroscopy. The x-ray fluorescence spectrometric method is an excellent multielemental techniuqe if simple preparation methods for x-ray samples are available. The technique has been used for the determination of arsenic in rocks[113], organically rich soils[114], river and sea sediments[115], air filters[116], and many other complex matrices. Although this method has seldom been applied to water analysis because of its inadequate sensitivity, a few attempts have been made to improve its relative sensitivity by employing isolation and preconcentration techniques. These techniques[117-124] include evaporation, precipitation, ion-exchange, and solvent extraction. Marcie[117] has developed a quantitative microdetermination method for arsenic, as well as some other trace elements, where the element reacts with ammonium pyrrollidine dithiocarbamate and is extracted into $CHCl_3$. The $CHCl_3$ extract is evaporated onto a filter paper disk and detected by x-ray fluorescence down to 0.005 ppm, with precision and accuracy of ±10%. Watanabe, Berman and Russell[124] determined arsenic in water via its precipitation, with a carrier (Ni^{+2}), by diethyldithiocarbamate and filtration through a Millipore filter. The precipitate accumulated on the filter is subjected to direct x-ray fluorescence using the K_α 48·78° (2θ) line. Spectral interference was reported only from the $L_{\alpha 1}$ line of lead.

5. Atomic Spectroscopy. Spectrometric determinations of arsenic are troubled by scattering and molecular absorption of the light (below 200 nm) by air, matrix species present in the combustion cell and by optical components of the instrument. Thus, improvement in these techniques can generally be accomplished by either operating in an inert atmosphere (e.g. vacuum, N_2, Ar, He), or using more efficient spectrometric excitation sources such as electrode and electrodeless discharge plasmas. The following is a short summary of various spectrometric sources applied to the determination of arsenic:

A. Atomic emission spectrometry. Hall and Lorell[125] determined the arsenic content of coal samples, utilizing the physical enrichment process developed for the analysis of minerals[126,127]. The working concentration range was 25-400 ppm, which is not adequate for most environmental samples, and accuracy and precision were 5-10%. Flame emission techniques are not sensitive enough either, due to the inadequate excitation energies provided by flames. On the other hand, plasma discharge spectrometric sources are much more efficient excitation sources and are therefore more sensitive. Dickinson and Fassel[128] have studied the detection of arsenic at its 228.8 nm spectral line with an RF plasma source. The system employs an ultrasonic nebulizer and desolvation before the solute enters the plasma, and can detect arsenic as low as 0.1 ppm. Kirkbright, Ward and West[129] have also studied arsenic emission with an RF plasma, but simplified the sample introduction technique by using an indirect nebulizer and expansion chamber which allowed the introduction of aqueous samples. The optical system was flushed with nitrogen, which made possible the monitoring of emission spectral lines in the 189-230 nm spectral region. Sensitivity was in the 0.1-0.3 ppm range, and few chemical and physical matrix interferences were found. Simpler plasma sources were also used in conjunction with highly volatile arsenic compounds preseparated from any complex matrix. Lichte and Skogerboe[49] using 2450 MHz microwave plasma determined generated arsine samples via their decomposition to free arsenic whose emission intensity was monitored. Detection limits better than 1 ng and an rsd of ±5% were obtained with good accuracy. Braman, Justen and Foreback[50] have similarly detected arsine in a d-c helium discharge with comparable sensitivity. Following its conversion to triphenyl arsine, arsenic was quantitatively determined with a detection limit of 15 pg, using the 2450 MHz microwave plasma interfaced to a gc[27]. Thus, it appears that plasma sources can serve as efficient ultra sensitive detectors for arsenic determination, using the atomic emission spectrometric mode.

B. Atomic absorption spectrometry. Arsenic has been determined by flame atomic absorption spectrometry in argon-hydrogen[130,131], air acetylene[132-134], and separated nitrous oxide-acetylene flames[135]. The detection limits reported are in the range of 0.5-1 ppm at the 193.7 nm As line. Extensive chemical interferences are observed from common ions when cool flames are used. The use of the separated N_2O-C_2H_2 flame overcomes most of these interferences. Thus flames, although most convenient sources, only partially fulfill the requirements of an ideal atomizer, namely (1) total sample utilization, (2) efficient and reproducible atomization, (3) low degree of ionization, (4) inert atmosphere, (5) long residence time to prolong observation time, (6) capability of atomizing solids, liquids and gases, (7) fast routine analysis, (8) low cost and simple operation. To increase the applicability range of arsenic determination by AA, other atomizers are used which attempt to better meet the above criteria. Ando, Suzuki, Fuwa and Valee[136] studied the applicability of the long tube (Vycor) absorption cell with argon, helium or nitrogen air-hydrogen flames. With this system, a long optical path was achieved with an almost inert atmosphere. As a result, sensitivities are improved while absorption by air was drastically reduced. Although sensitivity was 6 ng/ml, the technique suffers from multiple interferences; acid strength and concentration of salts and elements such as Ni, Co, Al are important parameters, and hence, careful control is required in the preparation of both standards and sample solutions. Chu, Barron and Baumgarner[10] determined the evolved arsine by sweeping it into an electrically heated absorption tube with argon as carrier gas. Since no flame was employed, the background absorption was less than that of the argon/hydrogen entrained air flame method, and the sensitivity was doubled. The rsd for a 0.4 μg arsenic standard was 0.36% compared to 4.8% for the argon-hydrogen flame[51]. Gandrud and Skogerboe[137] demonstrated the potential capability of a hollow cathode discharge as a means for vaporizing and atomizing samples for AA measurement. A detection limit of 10 ng was obtained, rsd values were at the 5.1-16.2% range. The technique, however, was not applied to any "real" sample. Spielholtz, Toralballa and Steinberg[12] compared the tantalum sampling boat and application flame AA techniques. The sampling boat requires only a fraction of a ml sample, whereas the aspiration method requires a minimum of 10 ml. The sensitivity obtained with the boat, 0.1 μg, was at least 5 times better than with the aspiration. Arsenic in coal and insecticides was determined by both methods.

Among the highly publicized AA techniques in recent years are the graphite and tantalum resistor heated atomizers. These devices use very small samples (1-50 μl) and generally provide excellent sensitivities for many elements. It is rarely, though, in AA spectroscopy that one gets something for nothing, and it will be worthwhile to pay attention to some of the shortcomings characteristic of these devices. First, the spectacular absolute sensitivities obtained become a little less attractive when they are expressed in terms of relative sensitivity, since only 1-50 μl aliquots can be analyzed. Second, the sensitivity and reproducibility depend strongly on the volume of the samples and the variation in their location inside the atomization chamber. Also, the performance of the atomizer tube usually deteriorates continuously, especially when highly salted samples are analyzed. Finally, the very small volume to which the vaporized sample is confined (which is the main cause for the excellent sensitivities obtained) results in very high non-specific spectral interferences compared to flame.

Nevertheless, these techniques, if used cautiously and with thorough pretreatment of the sample, can be rather useful, as demonstrated by Baird, Pourian and Gabrielian[4] who applied a carbon rod atomizer to the determination of arsenic in waste water. Their procedure requires the digestion of the sample and its preconcentration (8:1). Also, large quantities of Cl^- had to be removed by precipitation with Ag^+ and filtration, and various cations, by passage of the solution through a cation exchange column. Relative sensitivities of 10 μg/l were thus obtained with an rsd of 8%.

Knudson and Christian[138] compared the detection of arsine by an argon/air-H_2 flame to that by a graphite resistance furnace.

Detection limits with the furnace were improved approximately by one order of magnitude. Sensitivity was 1 ng and relative sensitivity 0.005 ppb. In this application, the furnace is used only as an atomizer, and no interfering matrix is present.

Chromatographic Methods

Total Arsenic. Von Endt, Kearney and Kaufman[139] separated arsenite, arsenate and monomethyl arsonic acid (MMAA) by thin layer chromatography. McBride and Wolfe[140] identified MMAA by thin layer electrophoresis and chromatography with radioautographic detection. Sachs, Michael, Anastasia and Wells[141] separated arsenite, arsenate, MMAA at dimethylarsinic acid (DMAA) by paper chromatography.

The detection of inorganic arsenic by gas chromatography (gc) via its transformation to thermally stable volatile derivatives has been studied by a few workers. Tadmor[142] and Vranti-Piscou, Kontoyannakos and Parissakis[143] have described the separation of arsenic tri-chloride, and Juvet and Fisher[144] that of arsenic trifluoride. Neither demonstrated any application of their method. Butts and Rainey[145] utilized their gc-ms system for the separation and detection of arsenite and arsenate, following their conversion to volatile trimethylsilyl derivatives. Schwedt and Rüssel[146,147] determined arsenic by gc following its conversion to triphenylarsine. The analytical procedure involved combustion of the sample in a Schöniger flask and absorption of the arsenic in hydrochloric acid solution, extraction of the arsenic as dithiocarbamate, and its phenylation with diphenylmagnesium after solvent evaporation. The triphenylarsine formed was separated by gc with detection limit of 0.4 ng and relative sensitivity (in sample) of 2 ppm.

A modification of this method was developed in our laboratory[27] eliminating the need for the tedious and uncertain vaporization step, and extending the relative sensitivity to 30 ppb for solid samples and 0.05 ppb for water samples. The analytical procedure involves the wet HNO_3-$HClO_4$ digestion of the sample, coprecipitation of arsenic with thionalide and filtration on a 0.8 μm Millipore filter, the transfer of the precipitate to a 20 ml vial, and its reaction with 1.5 M phenylmagnesium bromide-ether solution following the dissolution of the precipitate in benzene. The phenylation process is complete after 40 minutes, and excess phenylmagnesium bromide is destroyed with 2% thioglycolic acid-water solution. Aliquots of the organic layer are injected into the gc column, and the separated triphenylarsine is detected with a microwave emission spectrometric detector (MES)[27,148-151] via monitoring the emission intensity at the arsenic 228.8 nm spectral line. Sharp, well resolved peaks are obtained with a retention time in the 70-80 second range. Due to the excellent spectroscopic selectivity of the detector, and since the plasma is ignited after the solvent peak is eluted, only triphenylarsine is detected and no cleanup procedures are required for the extract.

The gc column is made of graphitized carbon beads (produced in our laboratory) coated with 1.5% FFAP. The close-to-perfect spherical shape of the beads and their inertness eliminated adsorption and tailing problems that were found for arsenic compounds with many other columns. Using the above method, coal, fly ash, orchard leaves, bovine liver, sea and river water were analyzed with accuracy and precision of 5-10%. Figures 1 and 2 show some typical chromatograms obtained.

Speciation-Organo-Arsenic. Determination of monomethylarsenic acid (MMAA) and dimethylarsinic acid (DMAA) is usually accomplished through the analysis of their total arsenic content. Among total arsenic methods used for this purpose are colorimetry[152-154], ion exchange chromatography[155], coulometry[156,157], NAA[158], titrimetry[159], and distillation[160]. Radioactive tracers have been used to study the transportation, persistence and degradation of herbicides in soils[161,162]. Nevertheless, there is a real need for analytical methods capable of separating and quantitatively determining the various organoarsenic compounds introduced into or produced by the environment.

Braman and Foreback[162] reduce arsenic compounds to their corresponding arsine derivatives with $NaBH_4$, accumulate them in a cold trap and selectively volatilize them into a dc helium plasma where they are spectrometrically detected down to 1 ng. The arsines are volatilized in the order of their boiling points, thus achieving a readout formally analogous to a thermal volatilization curve. Using this method, As(III), As(V), MMAA and DMAA were quantitatively determined in water, urine, bird eggshells, seashells and limestone at sub-ppb concentrations. Cox and Alexander[163] accumulated air samples above solutions of methylating micro-organisms and separated the generated trimethylarsine by gas chromatography. Johnson, Gerhardt and Aue[164] described two methods for the separation of MMAA by gc. The first involved the synthesis of the trimethylsilylate derivative of MMAA, which was highly volatile and chromatographed well. Unfortunately, the method was nonselective, since several arsenic compounds, including As_2O_3, have yielded the same product. In the second method, MMAA reacts with ethylene glycol[165] to form a selective derivative which chromatographs well. The derivative was detected either by a microwave plasma detector (1.5ng, detection limit) or by an alkali flame ionization detector (1 ng, detection limit). Lodmell[166] reacted MMAA and DMAA with allylthiourea[167] in aqueous solution at 70°C for 5 minutes. Aliquots were separated by gc and determined simultaneously by flame ionization detector (FID) and a N_2O/C_2H_2 flame spectrometric detector (FSD). Although the FSD is highly superior in selectivity (Figure 3), its sensitivity is inferior to that of the FID. We believe that the microwave emission spectrometric detector (MES) will furnish both the selectivity and sensitivity required for this analysis. The gc-MES system described earlier[27] has also been applied to the determination of various arsenic acids[27]. Although the study is incomplete, the preliminary results are very promising. The method is based on the reduction of the acids with $NaBH_4$ to their corresponding arsines and their separation by a gc column. The arsines

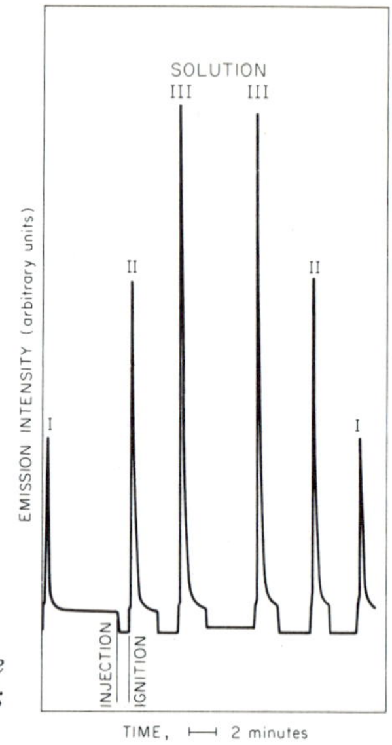

Figure 1. Chromatogram of triphenylarsine benzene solutions. I: 0.1 μg/ml; II: 0.2 μg/ml; III: 0.3 μg/ml (all injections are 3 μl).

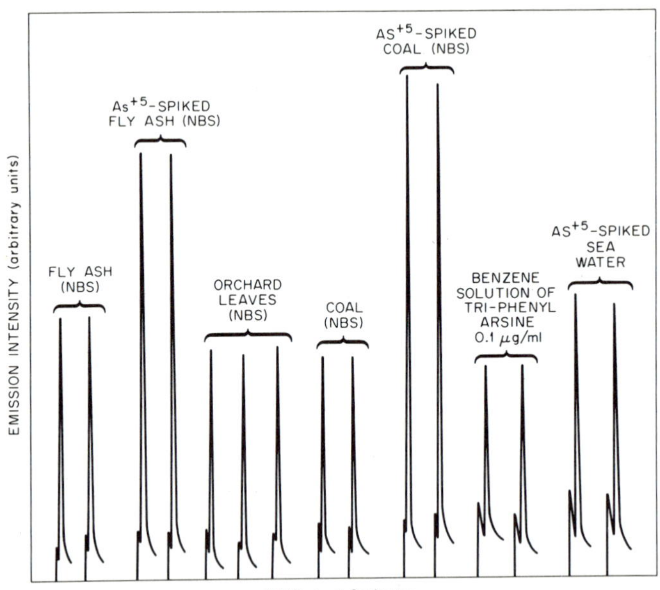

Figure 2. Determination of Total Arsenic as Triphenylarsine in Fly Ash, Orchard Leaves, Coal and Sb-Spiked Sea Water

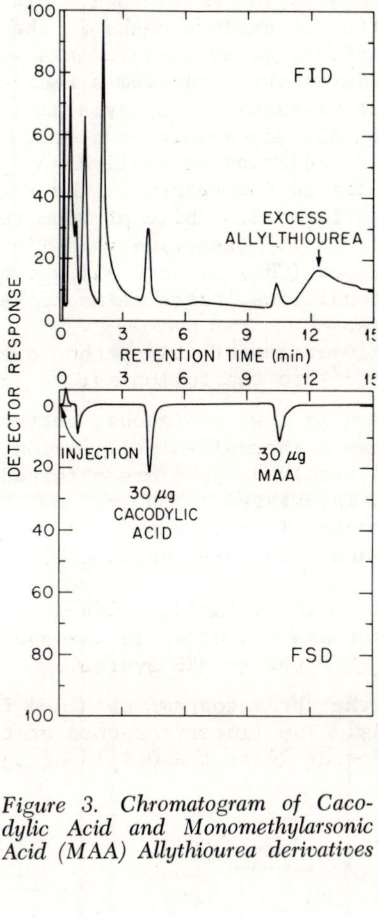

Figure 3. Chromatogram of Cacodylic Acid and Monomethylarsonic Acid (MAA) Allythiourea derivatives

generated are extracted into a benzene layer which is in direct contact with the aqueous reaction solution. In the future, the arsines will be accumulated on a solid adsorbant or a cold trap prior to their introduction into the column. The column is a 6% carbowax 20 M on 60/80 mesh chromosorb 101. As with triphenylarsine, the arsines are quantitatively detected via monitoring the arsenic emission intensity in the microwave plasma. To ensure the positive identification of various peaks in the chromatogram, the column was also interfaced to an organic mass spectrometer. Figure 4 shows the chromatogram obtained from a reduced mixture of arsenous oxide, monomethylarsonic, dimethylarsinic, ethylarsonic and propylarsonic acids. All acids were reduced to the expected corresponding arsines, in addition to small amounts of AsH_3, probably due to As_2O_3 impurities in the reagent. The reduction of cacodylic acid was less straightforward. While at room temperature, only $(CH_3)_2AsH$ was formed; at 75° (reaction vessel's temperature) AsH_3, $(CH_3)AsH_2$, $(CH_3)_2AsH$ and $(CH_3)_3As$ were all shown in the chromatogram. Under these conditions, there seems to be some ligand-exchange side-reactions.

The potential advantages of this method over the selective volatilization scheme[32] are the following:

1. gc separation is less ambiguous, better controlled and more reproducible than a thermal volatilization procedure.
2. Sharp, well resolved peaks are obtained and thus peak heights, rather than integrated areas are used for measurements.
3. The MES detector is more sensitive, detection limit approximately 30 pg, and much more reproducible than the dc discharge.
4. In contrast to the volatilization-dc plasma system, where severe deposition problems may occur in the spectrometric cell, no such problems occur with the gc-MES system.

Figure 5 shows the chromatograms obtained for 0.1 ppm benzene solutions of $(CH_3)_3As$. The linear response of the calibration curve was observed for at least the 0.1-10 ng range.

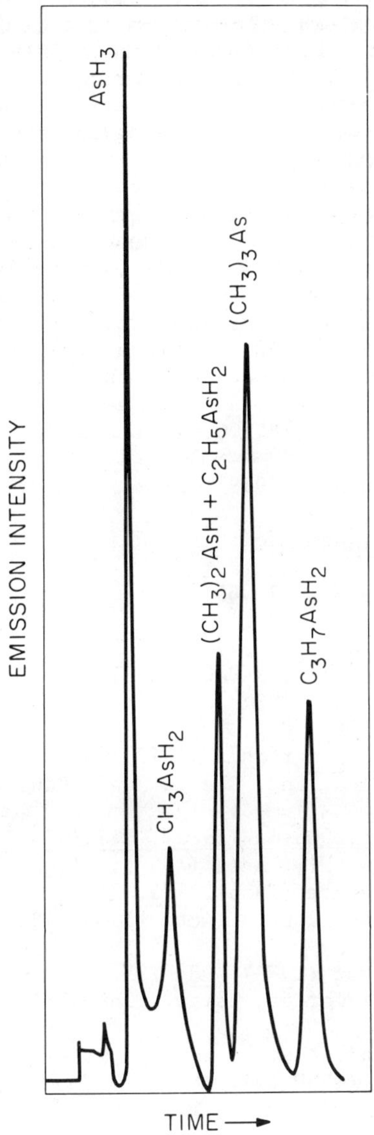

Figure 4. Chromatogram of alkylarsines

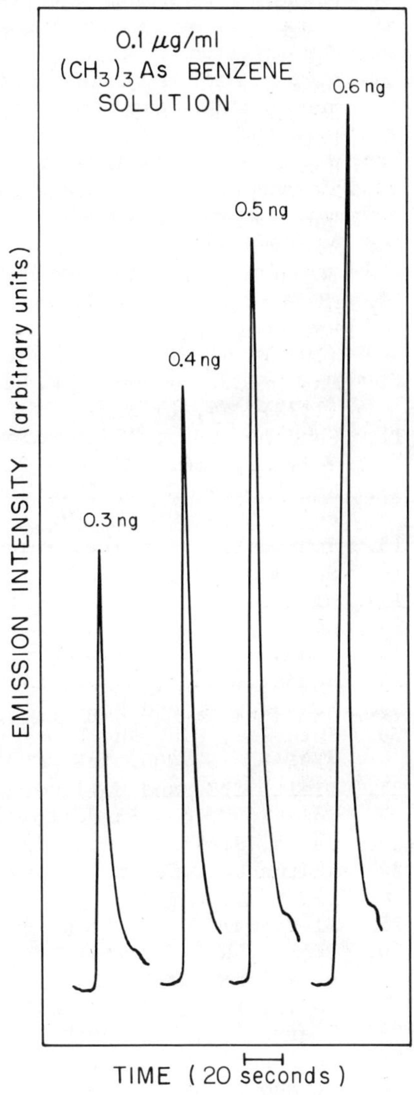

Figure 5. Chromatograms of trimethylarsine benzene solution

Literature Cited

1. Portmann, J.E. and Riley, J.P., Anal.Chim.Acta (1964) 31, 509.
2. Kingsley, G.R. and Schaffert, R.R., Anal. Chem.(1951) 23, 914.
3. Abernethy, R.F. and Gibson, F.H., U.S. Bur.Mines (1968) RI 7184.
4. Baird, R.B., Pourian, S. and Gabrielian, S.M., 166th Natl.Mtg. ACS, August 1973. Div. Environ. Chem. Preprints, 13, No.2, Paper 15.
5. Hamme, N.A., Young, A.L. and Hunter, J.A., Tech. Rept. (1970) AFATL-TR-70-107, Air Force Armament Lab., Elgin Air Force Base, Florida.
6. Adrian, W.J., Atomic.Abs.Newslett. (1971), 10, 96.
7. Bouchard, A., Atom.Abs.Newslett. (1973), 12, 115.
8. MacGee, J., J.Gas.Chromatog. (1968), 6, 48.
9. Jackson, A.J., Michael, L.M. and Schumacher, H.J., Anal.Chem. (1972), 44, 1064.
10. Chu, R.C., Barron, G.P. and Baumgarner, P.A.Q., Anal. Chem. (1972), 44, 1476.
11. Sandell, E.B., "Coloumetric Analysis", (1959) 3rd ed., Interscience, NYC.
12. Spielholtz, G.I., Toralballa, G.C. and Steinberg, R.J., Mikrochim. Acta (Wien) (1971), 918.
13. Schramel, P., Anal. Chim. Acta (1973), 67, 69.
14. Schramel, P., Anal. Chem. (1967), 39, 1480.
15. Migault, W., Chemiker Zeitung (1910), 34, 337.
16. George, G.M., Frahn, L.J. and McDonnell, JAOAC (1973), 56, 793.
17. Schöniger, W., Mikrochim. Acta (1955), 123.
18. Schöniger, W., Mikrochim. Acta (1956), 869.
19. Schwedt, G. and Rüssel, H.A., Chromatographia (1972), 5, 242.
20. Saterlee, H.S. and Blodgett, G., Ind. Eng. Chem., Anal. Edn. (1944), 16, 400.
21. Gleit, C.E. and Holland, A.D., Anal. Chem. (1962), 34, 1454.
22. Onishi, H. and Sandell, E.B., Mikrochim. Acta (1953), 34.
23. James, J.A. and Richards, D.H., Nature (1955), 175, 769.
24. Whitnack, G.C. and Brophy, R.G., Anal. Chim. Acta (1969), 48, 123.
25. Al-Sibbai, A.A., Fogg, A.G., Analyst (1973), 98, 732.
26. Braman, R.S., University of South Florida, Tampa, Fla., Private Communication.
27. Talmi, Y. (to be submitted to Anal. Chem.)
28. Pierrucini, R., Spectrochim. Acta (1950), 4, 189.
29. Ishibashi, M., Shigematsu, T., and Nishikawa, Y., Records Oceanogr. Works Japan (1960), 5.
30. Sugawara, K., Paper in Regional Symposium on Phys. Oceanogr., Oct.19-22 (1955).
31. Plotnikov, V.I. and Usatova, L.P., Zhur.Anal.Khim. (1964) 19, 1183.
32. Reichel, W. and Bleakley, B.G., Anal.Chem. (1974), 46, 59.

33. Santoliquido, P.M., Radiochem Radioanal. Lett. (1973), 15, 373.
34. Hanna, Z.G., Z. Anal. Chemie. (1965) 210 (5), 339.
35. Sergeeva, V.V., Levin, J.J., Tishchenko, L.I. and Dankova, V. Zhur. Anal. Khim. (1973), 27, 2188.
36. Klein, A.K. and Vorhes, F.A., JAOAC (1939), 22, 121.
37. Klein, A.K. and Vorhes, F.A. JAOAC (1942), 25, 284.
38. Wyatt, P.F., Analyst (1953), 78, 656.
39. Wyatt, P.F., Analyst (1955), 80, 368.
40. Velker, T.J., Anal. Chem. (1962), 34, 87.
41. Malissa H., and Schöffmann, E., Mikrochim.Acta. (1955), 187.
42. Mulford, C.E., Atom. Abs. Newslett. (1966) 5, 88.
43. Busev, A.I. and Ivanutin, M.L., Trudy Kommiss, Analit.Khim. Akad. Nauk, U.S.S.R. (1960), 11, 172.
44. Gagliardi, E. and Wöss, H.P., Anal.Chim.Acta (1969), 48, 107.
45. Byrne, A.R., Anal. Chim. Acta (1972), 59, 91.
46. Madsen, R.E., Atom. Abs. Newslett. (1971), 10, 57.
47. Fernandez, F.J. and Manning, D.C., Atom.Abs.Newslett., (1971) 10, 86.
48. Pollock, E.N. and West, S.J., Atom. Abs. Newslett. (1973) 12, 6.
49. Lichte, F.E. and Skogerboe, R.K., Anal. Chem. (1972) 44, 1480.
50. Braman, R.S., Justen, L.L. and Foreback, C.C., Anal.Chem. (1972), 44, 2195.
51. Fernandez, F.J., Atom. Abs. Newslett. (1973), 12, 93.
52. Ryan, D.E. and Lutwick, G.D., Can. J. Chem. (1953) 31, 9.
53. Gastinger, E., Mikrochem. Acta. (1972), 4, 526.
54. Hundley, H.K. and Underwood, J.C., JAOAC (1970), 53, 1176.
55. Small, H.G. Jr., and McCants, C.B., Soil, Sci. Soc. Am. Proc. (1961), 25, 346.
56. Dénigés, G., Compt. Rend. (1920), 171, 802.
57. Troug, E. and Meyer, A.H., Ind.Eng.Chem.Anal.Ed. (1929) 1, 136.
58. Heron, A.E. and Rogers, D., Analyst (1946), 71, 414.
59. Morris, H.J. and Calvery, H.O., Ind.Eng.Chem.Anal.Ed. (1941), 13, 760.
60. Rodden, H.J., J. Res. Natl. Bur. Std. (1940), 24, 7.
61. Woolson, E.A., Axley, J.H. and Kearney, P.C., Soil, Sci. (1971), 111, 158.
62. Gorgy, S., Fox, D.L. and Rabestraw, N.W., J. Marine Res. (1948), 7, 22.
63. Sugawara, K. and Kanamori, S., Bull. Chem. Soc. Japan, (1964) 37, 1358.
64. Gertnere, M., Jansones, E., Timofeeva, S.K. and Perets, V., Izv. Aka. Nauk. Latv. SSR. Ser.Khim. (1972), 3, 299.
65. Marczenko, Z., Mojsmi, M. and Skibe, H., Chemia Anal. (1972) 17, 881.
66. Stará, V. and Starý, J., Talanta (1970), 17, 341.
67. Markova, L.V. and Maksimenko, T.S., Zh.Anal.Khim (1970) 25, 1620.
68. Johnson, D.L. and Pilson, M.E.Q., Anal.Chim.Acta., (1972) 58, 289.
69. Javanovic, D., Krstic-Bogdanovic, E., and Stojadinovic, L., Acta. Pharm. Jugosl. (1971), 21, 1.

70. Ratliff, C.R., Hall, F.F. and Culp, T.W., Clin. Chem. (1972) 18, 717.
71. Morozova, S.F. and Mezhueva, N.S., Gidroliz. Lesokhim. Prom. (1972), 7, 9.
72. Lahrman, E., Nietruch, F. and Drews, M., Gesundh. Ing. (1973) 94, 258.
73. Smales, A.A., Mapper, D. and Fouché, K.F., Rep. U.K. Atom. Energy Auth., AERE R-5254, 1967.
74. Kiesl, W., "Modern Trends in Activation Analysis," NBS special Publications, Dept. of Commerce (1969), No.312, Vol.I, 302.
75. Morrison, G.H., Gerard, J.T., Travesi, A., Cunie, R.L., Peterson, S.F. and Potter, N.M., Anal. Chem. (1969), 41, 1633.
76. Laul, J.C., Wechten, D.R., Schmidt-Bleck, F. and Lipschutz, M.E., J. Radioanal. Chem. (1970), 4, 241.
77. Steinnes, E., "Activation Analysis in Geochemistry and Cosmochemistry", 113, A.D. Brunfett and E. Steinnes, Universitetsforlagt, Oslo (1971).
78. Heydron, K. and Damsgaard, E., Talanta (1973), 20, 1.
79. Lunde, G., Acta Chim. Scand. (1973) 15, 377.
80. Lunde, G., J. Amer. Oil Chem. Soc. (1972) 49(1), 44.
81. Kroon, J.J. and Oas, A.A., Reactor Centrum Nederland, Petten (RCN-124) Aug. 1970.
82. Robertson, D.E. and Carpenter, R., Report D8RPT7, 1972.
83. Eckhoff, N.D., Clark, R.W., Anthony, H.D. and Gray, A.P., Toxicol. Appl. Pharmacol. (1973) 24(2), 178.
84. Lacroix, J.P., Steinier, J. and Dreze, P., Anal.Lett. (1973) 6(6), 565.
85. Lee, C.L., Kim, N.B., and Park, E., J. Korean Nuclear Soc. (1973) 5, 279.
86. Lunde, G., J. Sci. Fd. Agric. (1968) 19, 432.
87. Becker, D.A., LaFleur, P.D., Proceedings 5th Univ. Missouri Conf. on Trace Subst. in Environm. Health, Univ. Missouri, Columbia, Missouri, 1971, p.559.
88. Takeo, T. and Shibuya, M., J. Food Sci. Technol. (1972) 19(2), 91.
89. Maruyama, Y. and Komiya, K., Radioisotope (1973) 5, 279.
90. Ray, B.J. and Johnson, D.L., Anal. Chim. Acta (1972) 62(1),196.
91. Gohda, S., Bull, Chem. Soc. Japan (1972) 45(6), 1704.
92. Geisman, J.R., Carey, W.E., Gould, W.A. and Alban, E.K., J. Food Sci. (1969) 34(3), 295.
93. Tanner, J.T., Friedman, M.H. and Holloway, G.E., Anal.Chim. Acta (1973) 66(3), 456.
94. Guinn, V.P., Bellanca, S.C., U.S. Natl. Bur. Stds. Special Publ. No. 312(1), (1969), 93.
95. Steinnes, E., Analyst (1972) 97, 241.
96. Onishi, H., "Handbook of Geochemistry," pp.33-E-5 and 51-E-3, K.H. Wedepohl, Springer Verlag, Berlin, 1969.
97. Zeman, A., Růžička, J., Starý, J. and Klečková, E., Talanta (1964) 11, 1143.

98. Ružička, J. and Starý, J., Inst. Nucl. Res., Report 992/63, Rez., 1963.
99. Arnold, A., Davis, S. and Jordan, A.L., Analyst (1969) 94, 664.
100. Zeman, A., Stary, J. and Kratzer, K., Radiochem. Radioanal. Letters (1970) 4, 1.
101. Akhmedov. G., Zhdanov, A.K., Uzbekh. Khim. Zhur. (1971) 2, 9.
102. Issa, I.M., Hamdy, M.H. and Misbah, A.S., Mikrochem. J. (1972) 17, 480.
103. Jacob, T.J. and Nair, C.G.R., Talanta (1972) 19, 347.
104. Griepink, B.F.A. and Krijgsmann, W., Mikrochim. Acta (1968) 3, 574.
105. Sandhu, S.J., Pahil, S.S. and Sharma, K.D., Talanta (1973) 20, 329.
106. Arnold, J.P. and Johnson, R.M., Talanta (1969) 16, 1191.
107. Hagiwara, K. and Murase, T., Bunseki Kagaku (1960) 14, 757.
108. Trushina, L.F. and Kaplin, A.A., Zhur. Anal. Khim. (1970) 25, 1616.
109. Melabs Scientific Instruments, Palo Alto, Calif. Application Note 2A.
110. Meyers, D.J. and Osteryoung, J., Anal. Chem. (1973) 45, 267.
111. Bruckenstein, S. and Johnson, D.C., Anal. Chem. (1964) 36, 2186.
112. Simon, R.K., Christian, G.D., and Purdy, W.C., Amer. J. Clin. Path., (1968), 49, 207.
113. Fabbi, B.P. and Espos, L.F., U.S. Geol. Surv. Prof. Pap. No. 800-B, P.147 (1972).
114. Leake, R.C. and Peachey, D., Inst. Mining Met. Trans., Sect. B (1973) 82, 25.
115. Tanaka, H., Moriguchi, Y., Yamamoto T. and Hashizume, G., Bunseki Kagaku (1972) 21, 1456.
116. Cate, J.L., Univ. Calif. Radiation Lab. Report UCRL 51038 (Rev. 1) Nov. 15, 1971.
117. Marcie, F.J., Envir. Sci. Tech. (1967) 1, 164.
118. Morris, A.W., Anal. Chim. Acta (1968) 42, 397.
119. Haberer, K., Vom. Wasser (1965) 17, 128.
120. Natelson, S., Leighton, D.R. and Calas, O., Mikrochem. J. (1962) 6, 539.
121. Tackett, S.L. and Brocious, M.A., Anal. Lett. (1969) 2, 649.
122. Riley, J.P. and Taylor, D., Anal. Chim. Acta (1968) 40, 479.
123. Watanabe, H., Berman, S. and Russel, D.S., Talanta (1972) 19, 1363.
124. Taylor, H. and Beamish, F.E., Talanta (1968) 15, 1497.
125. Hall, R.H. and Lovell, H.L. Anal. Chem. (1958) 30, 1665.
126. Shaw, D.M. Jvensu, O.L., and Ahrens, L.H., Spectrochim. Acta (1950) 4, 233.
127. Wedepohl, K.H., Geochim, et Cosmochim. Acta (1953) 3, 93.
128. Dickinson, G.W. and Fassel, V.A., Anal. Chem. (1969) 41, 21.
129. Kirkbright, G.F., Ward, A.F. and West, T.S., Anal. Chim. Acta (1973) 64, 353.
130. Kahn, H.J. and Schallis, J.E., Atom. Absorption Newsletter (1968) 7, 5.
131. Johns, P., Spectrovision (1970) 24, 6.

132. Allen, J.E., Spectrochim. Acta (1962) 18, 259.
133. Slavin, W. and Sprague, S., Atom.Abs.Newslett.(1964) 18, 1.
134. Chakrabarti, C.L., Anal.Chim.Acta (1968) 42, 379.
135. Kirkbright, G.F. and Ranson, L., Anal.Chem. (1971) 43, 1238.
136. Ando, A., Suzuki, M., Fuwa, K. and Valee, B.L., Anal. Chem. (1969) 41, 1974.
137. Gandrud, B.W. and Skogerboe, R.K., Appl. Spec. (1971) 25, 243.
138. Knudson, E.J. and Christian, G.D., Anal.Lett (1973), 6(12), 1039.
139. Von Endt, D.W., Kearney, P.C., and Kaufman, A.A., Agricultural and Food Chem. (1968) 16, 17.
140. McBride, B.C., Wolfe, R.S., Biochemistry (1971) 10, 4312.
141. Sachs, R.M., Michael, J.L., Anastasia, F.B. and Wells, R.M., Weed Sci. (1971) 19, 412.
142. Tadmor, J., J. Gas Chromatog. (1964) 2, 385.
143. Vranti-Piscou, D., Kontoyannakos, J., and Parissakis, G., J. Chromatogr. Sci. (1971) 9, 499.
144. Juvet, R.S.Jr. and Fisher, R.L., Anal.Chem. (1965) 37, 1752.
145. Butts, W.C. and Rainey, W.T. Jr., Anal.Chem. (1971) 43, 1538.
146. Schwedt, G. and Russel, H.A., Chromatographia (1972) 5, 242.
147. Schwedt, G. and Russel, H.A., Z.Anal.Chem. (1973) 264, 301.
148. McCormack, A.J., Tong, S.C. and Cooke, W.D., Anal.Chem. (1965) 37, 1470.
149. Bache, C.A. and Lisk, D.J., Anal. Chem. (1966) 38, 1757.
150. Bache, C.A. and Lisk, D.J., Anal. Chem. (1971) 43, 1950.
151. Talmi, Y., In press, Anal. Chim. Acta.
152. Chaney, A.L. and Magnuseon, H.J., Ind. Eng. Chem., Anal.Edn. (1940) 12, 691.
153. Saito, J., Nagaku Seisan Gigutsu (1962) 7, 8.
154. Small, H.G. Jr. and McCants, C.B., Soil Sci. Soc. Amer. Proc. (1961) 25, 1346.
155. Bogliano, G., Grassini, G., Ledevy, M. and Ossincini, L., J. Chromatog. (1964) 14, 238.
156. Bruckenstein, S., Johnson, D.C., Anal. Chem. (1964) 36, 2186.
157. Simon, R.K., Christian, G.D., Pardy, W.C., Am.J.Clin.Path. (1968) 49, 207.
158. Lunde, G., Int. Rev. Gesamten Hydrobiologie (1967) 52, 265.
159. Bassett, J., Analyst (1963) 88, 238.
160. Stone, R.L., J. Assoc. Off. Anal. Chem. (1967) 50, 1361.
161. Woolson, E.A. and Kearney, P.C., Environ.Sci.& Tech. (1973) 7(1), 47.
162. Braman, R.S. and Foreback, C.C., Science (1973) 182, 1247.
163. Cox, D.P. and Alexander, M., Bull. Envir. Contam. and Toxicol. (1973) 9(2) 84.
164. Johnson, L.D., Gerhardt, K.L. and Aue, W.A., Sci. Total Environ. (1972) 1, 108.
165. Salmi, E.J., Merivuori and Laaksonen, E., Suom. Kemistilehti (1946) B19, 102.
166. Lodmell, J.D., Ph.D. Dissertation, Univ. of Tenn. 1973.
167. Raiziss, G.W., and Gavron, J.L., "Organic Arsenical Compounds" Chemical Catalog Co., p.81 (1923).

3

Behavior and Phytotoxicity of Inorganic Arsenicals in Soils

L. M. WALSH and D. R. KEENEY

Department of Soil Science, University of Wisconsin, Madison, Wisc. 53706

Introduction

Inorganic arsenicals have been used in agriculture as a pesticide or plant defoliant for many years. In localized areas oxides also contaminate soils by fallout from smelting ore, especially sulfide ores, and from burning coal. Past use of arsenicals has not resulted in serious problems with human or animal poisoning but severely reduced plant growth due to As phytotoxicity has been observed, especially on former orchards and cotton fields. Once toxicity is observed, it persists for several years even though no additional As is added. Most attempts to ameliorate As toxicity by the addition of salts which might reduce the level of available As in soils have not been very successful.

Historically, inorganic arsenicals (lead arsenate, calcium arsenate) were used as insecticides beginning in the latter part of the 17th century (1). Paris Green (which contains cuprous arsenite) was successfully used to control the Colorado Potato Beetle in eastern United States by 1867 (2). Lead arsenate has been used for insect control on a variety of fruit trees and, for many years, was the only insecticide that would successfully control codling moth in apple orchards and horn worm on tobacco. Calcium arsenates have been applied to cotton and tobacco fields to protect these crops from boll weevils, beetles and other insects. Arsenic trioxide has been widely used as a soil sterilant (3) and sodium arsenite has been used for aquatic weed control (4) and as a defoliant to kill potato vines prior to tuber harvest (5).

In recent years use of inorganic arsenicals has decreased markedly. Sodium arsenite has been banned and can no longer be used as a vine-killer or defoliant (6). Organic arsenicals have largely replaced inorganic As salts as herbicides and since they are applied at a lower rate of application, this has reduced the amount of As applied to the soil. Also, much lower quantities of lead arsenate are now used in orchards because fruit growers rely primarily on carbamates and organic phosphates to control insects. However, where long-term residual insecticidal activity is needed to protect orchards from chewing insects, lead arsenate is still being used to a limited extent by some growers. Also, inorganic arsenicals continue to be used on golf greens and fairways in some areas of the United States to control annual bluegrass.

The primary objectives of this paper are: (a) to review the relationship between arsenic sorption and soil properties and the possible mechanisms of As sorption in soils, (b) to review various soil test methods in terms of their ability to predict As uptake and phytotoxicity, (c) to review relationships between soil As and plant uptake which result in phytotoxicity and potential hazards to humans or animals consuming plants grown on soils containing high levels of As, and (d) to review methods of ameliorating As toxicity.

Occurrence and Distribution of As in Soils

Arsenic is ubiquitous in nature and is found in detectable amounts in nearly all soils and in many rocks and minerals. Surveys of soils in the United States show that As levels range from 0.2 to 40 ppm for uncontaminated soils, and up to 550 ppm for As-treated soils (7, 8). Comparisons of As-treated vs. untreated soils were made by several researchers and are reported in Table 1. These data indicate that untreated soils seldom contain more than 10 ppm As. In fact, Fleischer (9) reports that the average As content of untreated soils is only 5 ppm. Even though soils can vary greatly in their native As content, the levels of As in noncontaminated soils generally have not been high enough to cause phytotoxicity and they do not represent a significant health hazard (10). The As content of most rocks and minerals is similar to that found in soils, except for sulfide ores, sedimentary Fe and Mn ores, and rock phosphate which occasionally contain as much as 2000 ppm As (9). Coal usually contains about 25 ppm of As (11). Even though soils and soil parent materials vary considerably in their native content of As, no clearly defined relationship has been noted between the As content of soils and the climatic conditions or geological formations in which the soils were developed.

Under field conditions high levels of total soil As have been observed more often where crops have been treated with an arsenical insecticide as compared to a defoliant. This is due to the difference in the total amount of As applied. For instance, soil analysis indicates that some orchards have received over 1000 kg As/ha (12) as compared to about 10 kg As/ha where sodium arsenite was used as a potato vine defoliant (13).

Fate of Arsenic in Soils

Arsenic Chemistry. While As chemistry is in many ways similar to the other Group V elements, it is more metallic (i.e. more labile in behavior) than P, which is essentially covalent. In

Table 1. A comparison of As levels in As-treated and uncontaminated soils in North America

Sampling site	Total As Content: Uncontaminated soil	Total As Content: Treated soil*	Crop	Source of data
	ppm As			
Colorado	1.3-2.3	13-69	orchard	(61)
Florida	8	18-28	potato	(12)
Idaho	0-10	138-204	orchard	(12)
Indiana	2-4	56-250	orchard	(12)
Maine	9	10-40	blueberry	(12)
Maryland	19-41	21-238	orchard	(8)
New Jersey	10.0	92-270	orchard	(12)
New York	3-12	90-625	orchard	(12)
North Carolina	4	1-5	tobacco	(30)
Nova Scotia	0-7.9	10-124	orchard	(28)
Ontario	1.1-8.6	10-121	orchard	(62)
Oregon	2.9-14.0	17-439	orchard	(50)
	3-32	4-103	orchard	(12)
Washington	6-13	106-830	orchard	(34)
	8-80	106-2553	orchard	(12)
	4-13	48	orchard	(44)
Wisconsin	2.2	6-26	potato	(54)

* These are results from soils that had been repeatedly treated with an As pesticide or defoliant. Soils treated experimentally are not included.

strongly reducing environments, elemental As and arsine (-3) can exist, but arsenate (+5) is the stable oxidation state in oxygenated environments. Under moderately reduced conditions such as flooded soils, arsenite (+3) may be the dominant form (14). Evidence exists (15) that dimethylarsinic acid may be an important and ubiquitous As compound which has not heretofore been identified due to lack of methods for detection and quantification.

Biological Transformations. The -3, +3, and +5 valence states can form compounds containing the C-As bond and are readily interconverted by microbes, although there is no evidence the energy produced by oxidation is used for microbial growth (16, 17). Fleisher (9) and Wood (16) have suggested that microbial formation of volatile arsine or other volatile reduced As compounds might play a role in the discharge of As to the atmosphere. Reed and Sturgis (18) indicated that arsine may be lost from flooded soils. Wood (16) pointed out the marked similarities between the Hg and As cycles. According to Wood, in reduced environments such as sediments, arsenate is reduced to arsenite and methylated to methylarsenic acid or dimethylarsinic acid. These compounds, which are extremely toxic, may be further methylated (trimethylarsine) or reduced (dimethylarsine) and may volatilize to the atmosphere where oxidation reactions will result in reformation of dimethylarsinic acid. Thus studies which do not monitor the various possible forms of As can result in misleading information.

While precise information of the oxidation-reduction potential at which arsine, arsenite, and arsenate dominate is not available, Turner (17) calculated the E_0' for the arsenate-arsenite reaction to be +77 to +167 mV (pH 7.0). He suggested that bacterial oxidation of arsenite involved an adaptive arsenite-dehydrogenase system. It is possible that this enzyme complex evolves as a protective mechanism. The optimum pH for oxidation was between 6.0 and 6.7, and 15 strains of heterotrophic bacteria capable of carrying out this reaction were readily isolated. Deuel and Swoboda (14) obtained experimental evidence that arsenite oxidation occurs at about +100 mV. This work shows that arsenite will be oxidized in aerated soils, where the oxidation-reduction potential commonly is in the range of +400 to 600 mV. However, the rate of arsenite oxidation in soils is also of importance in estimating the residual activity of arsenite. While rate data have not been obtained, the perfusion experiments of Quastel and Scholefield (19) showed complete oxidation of .0025*M* sodium arsenite in 8 days in an unenriched soil. When the soil was enriched by perfusion with sodium arsenite until constant oxidation rates were obtained, complete oxidation occurred in about 4 days. Thus arsenite will not persist for long periods in well aerated soils which have significant biological activity.

Sorption of Arsenic in Soil. In most soil systems, the chemistry of As becomes the chemistry of arsenate. This ion has properties similar to phosphate including formation of insoluble salts with a number of cations, and sorption by soil constituents. For a comprehensive discussion of the behavior of soil and fertilizer phosphorus, see Black (20) and references cited therein.One major difference between P and As appears to be that while soils contain appreciable organic-combined phosphate, they do not contain measurable levels of organic As (21).

Solubility reactions may play a role in As retention by soils. For example, iron arsenate is extremely insoluble, maintaining an arsenate concentration of only about $10^{-11}M$ in solution, compared to about $10^{-5}M$ for the Ca or Mg arsenates (22). The modified Chang and Jackson (23) procedure for fractionation of inorganic soil P was first applied to As-contaminated soil by Johnson and Hiltbold (21). They found that about 90% of the As was associated with the clay fraction of a Chesterfield sandy loam previously treated with organic arsenicals, and that most of this As was associated with soil Al. However, this approach assumes that As extracted by ammonium fluoride is Al-bound and that removed by sodium hydroxide is Fe-bound. This is not necessarily the case as it has been shown that some of the P released by ammonium fluoride is subsequently resorbed by other soil constituents and then removed by the sodium hydroxide extraction (24). This was confirmed for As by Jacobs et al. (25) who also applied the Williams et al. procedure (which corrects for P resorption) to As-contaminated soils. They obtained results suggesting that most of the As was sorbed by amorphous Fe and Al components in soils.

Woolson et al. (12) conducted an extensive survey of As forms in As-contaminated soils from throughout the United States. They found that, as would be predicted, soils with a high reactive Fe (oxalate extractable) content had predominately Fe-bound As, but that when the reactive Fe was low, the dominant form of As was controlled by the relative levels of exchangeable Ca and reactive Al (see also reference 26).

A number of reports have indicated that As is more toxic on coarse-textured (i.e., sandy) than on fine-textured (i.e., high in clay) soils (3, 27, 28, 29, 30, 31, 32, 33, 34, 35). This is explained in part by the marked increase in surface area as the clay content increases. Also, as would be predicted from the As fractionation work discussed previously, sorption of arsenate increases with the amount of "active" (extractable) Fe or Al content of soils (12, 25, 26, 30, 36, 37, 38).

Sorption of As by soils is time dependent. Indications of reversion to less soluble (and hence less toxic) forms of As have been obtained in the field (39), greenhouse (35), and laboratory (26, 40). This is illustrated by Figure 1 (from reference 25) which shows the amount of As extracted from a sand (Plainfield) and a silty clay loam (Waupun) after equilibration for up to 12 months. More As was extracted by the Bray P-1 ($0.025N$ hydrochloric acid + $0.3N$ ammonium fluoride) than by $1.0N$ ammonium acetate (pH 7.0) and more As was removed from the sand than the silty clay loam. From 3 to 6 months equilibration was required before As reached an apparent equilibrium with these soils. Similarly, Woolson et al. (26) found that soluble (ammonium acetate or ammonium chloride extractable) As decreased to a constant value after about 4 months, and that the rate of decrease differed among soils. Iron-bound As continued to form even after the Al-bound fraction had reached a maximum level and begun to decline. Decline in As toxicity with time in field and greenhouse studies have been reported (3, 27, 35). However, in a field study, Steevens et al. (13) found little decline in As phytotoxicity to peas or potatoes with time.

Arsenic Movement and Persistence in Soil

While downward movement (leaching) of phosphate in soils is believed to be very limited (20) several workers have indicated that arsenate leaching may be a significant factor in reducing As toxicity of surface soils (3, 13, 26, 33, 41). The profile distribution of As in a Plainfield sand which had received up to 720 kg/ha of As in 1967 was investigated by Steevens *et al*. (13). These results, along with 1974 profile analyses by the authors, are presented in Table 2.

Table 2. Profile distribution with time of total soil As in a Plainfield sand after surface application of sodium arsenite in 1967. [J. Environ Qual. (13)]

Year	As Applied kg/ha	Depth, cm 0-23	23-38	38-53	53-68	68-83	LSD .05
		ppm As					
1967	0	3.0	--	--	--	--	--
	45	11.0	--	--	--	--	--
	90	23.0	--	--	--	--	--
	180	73.0	--	--	--	--	--
	720	250.0	--	--	--	--	--
1968	0	3.0	--	--	--	--	--
	45	19.0	--	--	--	--	--
	90	26.0	--	--	--	--	--
	180	63.0	--	--	--	--	--
	720	150.0	--	--	--	--	--
1970	0	3.6	1.2	1.8	1.3	1.2	0.8
	45	14.1	2.3	1.4	1.9	1.6	1.1
	90	27.0	3.4	1.5	1.2	1.4	3.2
	180	45.0	4.8	1.9	1.7	1.1	4.5
	720	100.0	65.0	8.6	3.3	2.0	4.2
1974*	0	1.6	1.0	1.3	.5	.6	.9
	45	8.1	1.2	1.4	.7	.7	4.5
	90	14.4	1.8	1.6	1.3	1.3	5.1
	180	21.6	2.8	1.5	1.1	1.3	9.9
	720	82.8	50.6	19.6	2.6	3.7	24.9

* Walsh and Keeney, unpublished data, 1974.

Through 1970, these plots received about 75 cm of precipitation and 50 cm of irrigation per year. Alfalfa has been grown on the experimental area during the past 3 years, but it was not harvested. The plots were not irrigated since 1970. Phytotoxicity persisted at the 180 and 720 kg/ha As plots from 1967 to 1970, and the 720 kg/ha plots are still essentially barren in 1974, 7 years after As application on a coarse textured soil. However, there has been a significant decline in total As with time in the surface soil of all As-treated plots. Part of this decrease can be explained by As leaching; by 1970 As had definitely been leached to the 38-53 cm depth in the 720 kg/ha plots, and leaching rate was related to the amount applied in agreement with the work of Tammes and de Lint (33). By 1974, the surface soil of the high As-treated plots had decreased further, but the rate of decrease was much less than in previous years. Significant amounts of As had accumulated in the subsoil, and in the 720 kg/ha plots, total As was above background to 83 cm.

These results show that As will leach in a sandy profile, and that in time sufficient As will be leached to significantly reduce phytotoxicity. However, the rate of As loss from the surface soil decreased markedly with time, indicating that As is persistent even in a soil with low sorption capacity.

When the profile results are summed, all of the As applied in 1967 cannot be accounted for as total As in the profile. One possibility is that significant amounts of As are lost from this soil by wind erosion (40). Tammes and de Lint (33) calculated an average half life of 6.5 ± 0.4 years for As persistence on two Netherland soils. Since quantitative profile As balances on the plots reported in Table 2 have not been achieved, half-life calculations with these data are not possible. However, in the 720 kg/ha plots, total As in the surface soil had declined to 40% of the original value in only 3 years, and to 33% by 7 years.

Arsenic Phytotoxicity

Soil "unproductiveness" due to arsenical poisoning resulting from heavy application of arsenical pesticides has been noted by many researchers (3, 28, 42, 43, 44). In reviewing the literature, it is apparent that the most serious problems occur in fields where orchards are removed and then replanted to an agronomic or horticultural crop (28, 44). Substantial amounts of arsenical herbicides have been used on golf greens but As toxicity has not been reported on bent grass or coastal bermuda grass.

Arsenic is not an essential plant nutrient but occasionally small yield increases have been observed at low levels of As, especially for tolerant crops such as potatoes, corn, rye and wheat (40, 45, 46). Liebig *et al.* (47) reported that root growth of lemon plants in solution culture was enhanced by 1 ppm As as arsenate or arsenite; 5 ppm of either form of As was toxic and adversely affected both top and root growth. Woolson (46) postulated that As response was due to stimulation of plant systems by small amounts of As since other pesticides (e.g. 2, 4-D) stimulate plant growth at sublethal dose levels. Another reason for the yield increase might be that arsenate would displace phosphate from the soil with a resultant increase in phosphate availability (40).

Plant Sensitivity. Plants vary considerably in their tolerance to high levels of soil As. In his review on As, Liebig (43) included the following classification of vegetables and small fruits in terms of their tolerance to water-soluble As:

Very tolerant: Asparagus, potato, tomato, carrot, tobacco, dewberry, grape, and red raspberry.

Fairly tolerant: Strawberry and sweet corn, beet and squash.

Low or no tolerance: Snap bean, lima bean, onion, pea, cucumber, alfalfa and other legumes.

Liebig (43) also indicated that rye and Sudan grass were very tolerant to soluble As. Recently, Deuel and Swoboda (29) reported that soybeans were more sensitive to As toxicity than cotton. Jacobs *et al*. (40) found that crop tolerance to As was in the following order: potatoes > peas > sweet corn > snap beans. Woolson *et al*. (32) reported the As tolerance of six vegetable crops to be as follows: Cabbage > tomato > radish > spinach > lima beans > green beans.

Relationship of Extractable As to Plant Growth. Many researchers have attempted to relate the amount of As in the soil with plant growth. As would be expected total soil As is not a good predictor of water soluble As or As phytotoxicity when these relationships are compared among soils with widely differing characteristics (12, 40, 48). However, total As accurately predicts phytotoxicity when studies have been limited to a narrow range of experimental treated soils (40, 49). In general, a highly significant relationship has been observed between some form of extractable As and plant growth. Extractants commonly used to measure available As include hot water, Bray P-1, mixed acid (0.05*N* hydrochloric acid + 0.025*N* sulfuric acid), 0.5*N* hydrochloric acid, 1*N* ammonium chloride, 1*N* ammonium acetate (pH 7.0) and 0.5*N* sodium bicarbonate. The results of a few recent studies, summarized in Table 3, show that soil type and plant specie influence the "critical" level of soil As at which a yield depression might be expected. The "critical" level ranged from 25 to 85 ppm As and from 3 to 28 ppm As for total soil As and for water soluble As, respectively. The critical levels for available As as measured by the Bray P-1, the mixed acid or sodium bicarbonate ranged from 10 to 22 ppm As.

Greenhouse work by Woolson *et al*. (46) conclusively shows that plant growth on As-contaminated soils is related to total As and available As. All correlations presented in Figure 2 are highly significant, however, use of the available tests more effectively predicted plant response as compared to total As. The three available As tests compared in Figure 2 were equally effective in predicting response to As. In a related study Woolson (32) found the growth of six vegetable crops were reduced by 50% at levels of 6.2 to 48.3 ppm of available As as measured with the mixed acid procedure. These data presented in Table 4 also show that the available As levels account for 64 to 83% of the variation in yield for six greenhouse-grown vegetable crops.

Greenhouse experiments were also conducted by Deuel and Swoboda (29) in which experimentally treated soils were used to relate levels of As extracted from the soils with yields of cotton or soybeans. These results, presented in Table 5, show that

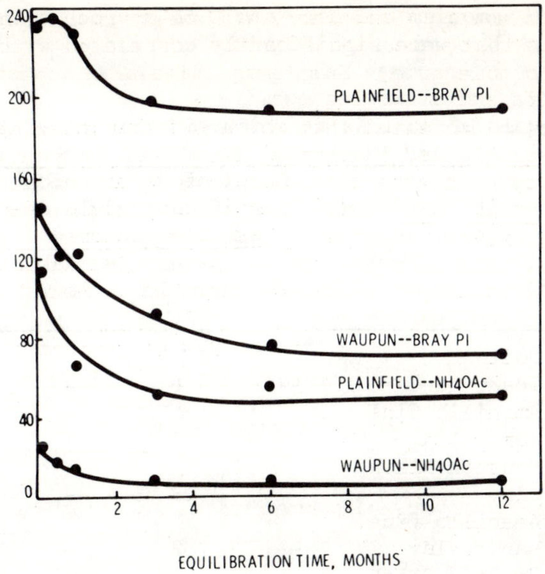

Soil Science Society of America, Proceedings

Figure 1. Extractability of added As (320 μg/g) from Plainfield sand and Waupun silty clay loam by Bray P-1 and NH_4OAc as related to equilibration time (25)

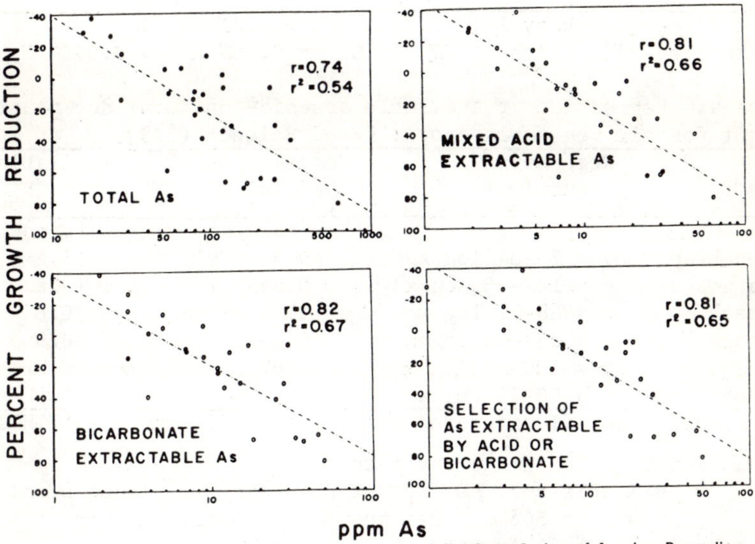

Soil Science Society of America, Proceedings

Figure 2. A comparison of four methods for the estimation of As availability by a correlation between growth reduction (4-week-old corn) and the log of As concentration in soils (46)

hot water, 1*N* ammonium chloride and 0.5*N* hydrochloric acid extracted As levels that were significantly correlated with plant growth.

Table 3. Levels of soil As at which significant yield depressions occur.

Crop name	Soil type	Level of As at which significant yield depressions occurred: Total As	Water soluble As	Available As	Source of data
		------ ppm As		---------	
Blueberry	Colton loamy sand	44	6	--	(49)
Cotton	Amarillo fine sandy loam	--	8	--	(29)
Cotton	Houston Black clay	--	28	--	(29)
Soybean	Amarillo fine sandy clay	--	3	--	(29)
Soybean	Houston Black clay	--	12	--	(29)
Potatoes, sweet corn	Plainfield loamy sand	68	--	22†	(40)
Snap beans, peas	Plainfield loamy sand	25	--	10†	(40)
Corn	(Avg. of 13 soils)	85	--	10‡	(46)

† Extracted with Bray P-1 (0.025*N* HCL + 0.3*N* NH_4F)

‡ Extracted with either 0.5*N* $NaHCO_3$ or 0.05*N* HCl + 0.025*N* H_2SO_4

Table 4. Regression of available arsenic* on total dry-plant weight for six vegetable crops. [Weed Science (32)]

Crop	Regression equation†	Correlation coefficient (r)	Available As at GR_{50}‡ (ppm)
Green beans	y = 77-34 log x§	0.89	6.2
Lima geans	y = 107-55 log x	0.83	10.9
Spinach	y = 88-37 log x	0.91	10.6
Cabbage	y = 114-38 log x	0.80	48.3
Tomato	y = 109-42 log x	0.87	25.4
Radish	y = 96-36 log x	0.81	19.0

* Available As-soil extracted with 0.05*N* H_2SO_4 and 0.025*N* HCl

† Calculated by least squares method

‡ GR_{50} = available soil As content necessary to reduce plant growth to 50% of that grown on non-arsenate-treated soil

§ x = available soil As

Table 5. Correlation coefficients for yield vs. extractable arsenic. [Journal of Environmental Quality (29)]

Soil	Crop	Extractant*		
		H_2O	HCl	NH_4Cl
Amarillo	Soybeans	-.943	-.915	-.914
Houston	Soybeans	-.968	-.931	-.938
Amarillo	Cotton	-.951	-.830	-.960
Houston	Cotton	-.954	-.895	-.918

* All of the correlation coefficients are significant at the 0.05 level.

Jacobs et al. (40) studied the relationship between soil As and the yields of several vegetable crops grown on an irrigated Plainfield Sand. All extractants were negatively correlated with crop yields at the 0.01 probability level (Table 6). Total As and ammonium acetate or Bray P-1 extractable As were equally effective in predicting reduced yields as levels of soil As increased.

Table 6. Relationships between total or extractable soil As and yield of vegetable crops. [Agronomy Journal (40)]

As fraction	Crop*			
	Potatoes	Peas	Snap beans	Sweet corn
NH_4OAc	-0.91	-0.85	-0.73	-0.91
Bray P-1	-0.91	-0.88	-0.77	-0.93
Total	-0.92	-0.87	-0.76	-0.03

* All correlation coefficients significant at the 0.01 probability level.

Prior work clearly indicates that several soil tests can be used to predict As phytotoxicity. Even though water soluble and total As satisfactorily predict As phytotoxicity, soil testing laboratories will find it more convenient to use Bray P-1, sodium bicarbonate or the mixed acid extractant because these extractants are now routinely used for available P. Furthermore, they extract more As than water, and eliminate the digestion step needed for total As.

Plant Uptake of Arsenic. Application of many chemical elements results in substantial increases in crop assimilation of those elements. In the case of As such bioaccumulation would be hazardous to human beings or animals because of toxicity of As and its possible relationship to cancer, artereosclerosis and chronic liver diseases (10). Fortunately, the edible portion of plants seldom accumulates a hazardous level of As, primarily because most plants are sensitive to As toxicity and growth is usually severely reduced before a level of As hazardous to man or animals accumulates in the plant. When As poisoning does occur, it is usually due to direct ingestion of a surface residue of As or the ingestion of spring water or muds containing abnormally high levels of As (50, 51).

The edible portion of most fruits and vegetables grown in As-treated soils contain less than the tolerance of 2.6 ppm as allowed by the U.S. Public Health service for As-treated fruits and vegetables. The highest levels of As are found in plant roots, the vegetative top growth is intermediate, and edible seeds and fruits contain the lowest level of As. Jones and Hatch (50) analyzed vegetable plants growing on As-treated soils on several experimental farms in Oregon and found the roots, tops (stems and leaves) and edible portion to contain an average of 7.1, 5.0, and 1.2 ppm As, respectively. Fleming et al. (52) and McLean et al. (53) observed that vegetables grown on soils treated with high levels of lead arsenate seldom contained more than 1 ppm As in the edible portions. A wide range of forage and vegetable crops were grown by Jones and Hatch (50) on As-treated and adjacent untreated soils in Oregon. They found the As content in the edible plant parts to be only 1.2 and .41 ppm As in the treated and untreated areas, respectively. On the other hand, Small and McCants (30) found that the concentration of As in flue-cured tobacco varied from 2 ppm As where no As was applied to 14.3 ppm where 54 kg As/ha or lead arsenate had been applied to the soil.

Arsenic levels of less than 1.0 ppm were found by Jacobs et al. (40) in potato tubers even where severe As toxicity occurred. The peeling, however, contained up to 48 ppm As where the soil had been experimentally treated with 720 kg As/ha. The authors suggested the high levels of As in the peelings was due to minute quantities of As-contaminated soil adhering to the surface of the potato tuber, rather than assimilation of As into the peeling itself. A survey of potato grower fields in Central Wisconsin where 8 to 65 kg As/ha as sodium arsenite had been applied over a period of several years revealed that the potato peelings did not contain more than 2-3 ppm As (54). It was concluded, thus, that past usage of sodium arsenite in Wisconsin had not caused harmful levels of As to accumulate in potato tubers. High As levels have also been observed in washed, unpeeled radishes (32). As with potatoes, part of the As may have been adsorbed on the surface of the radish root.

Woolson (32) found a significant positive relationship between available soil As and the concentration of As in the whole plant for six greenhouse grown vegetable crops (Table 7). Correlations between available soil As and the concentration of As in the edible plant parts were generally poorer, especially for plants in which the seed or fruit was consumed (32). As previously noted, plants tend to exclude As from the seeds and fruits. Hence, soil tests for available As would not be a reliable indicator of the amount of As in the edible plant tissue. Woolson (32) has pointed out that where the edible part of a crop is not the root or whole plant, it is unlikely that the tolerance level for As (2.6 ppm) would be exceeded even where As toxicity reduced growth by 50%.

Table 7. Regression of available soil As on the As content in whole dry plant or edible dry-plant part [Weed Sci. (32]

Crop	Regression equation	Correlation coefficient	Arsenic at GR_{50}
		(r)	(ppm)
Green bean	y = 0.4 + 4.2 log x*	0.93	3.7†
Lima bean	y = 0.5 + 1.2 log x	0.49	1.7
Spinach	log y = -0.129 + 1.1 log x	0.90	10.0
Cabbage	y = 0.4 + 1.8 log x	0.77	3.4
Tomato	y = -0.1 + 3.3 log x	0.80	4.5
Radish	log y = -0.147 + 1.4 log x	0.88	43.8

* x = available soil As

† Available As at GR_{50} was used to calculate As content (see table 4).

Wherever soils have been treated with substantial quantities of As, the surface of the aerial portion of the plant may be contaminated with dust. The work of Jacobs *et al.* (40) showed that the concentration of As in potato leaves, stems and petioles was related to the surface area of the plant parts and the prevailing wind in relation to plots treated with high rates of As. Plant As concentration bore little relationship to As treatment; thus, these workers concluded that substantial external contamination of the plant tissue in the field had occurred. In light of this work, it appears that some of the data in the literature on plant uptake of As would be questionable unless the researcher very carefully washed the plant tissue to remove adsorbed soil particles. For instance, Jones and Hatch (50) reported that the top growth of vegetable plants growing on untreated soil adjacent to As-treated soil contained 3.1 ppm As while the roots from these plants contained only 1.1 ppm As. Since roots normally accumulate more As than the top growth, it appears that the top growth must have been contaminated with soil particles from the adjacent orchard soil which had been treated with As for several years.

Alleviation of As Toxicity

Since registration of inorganic arsenicals for use on nearly all vegetable and agronomic crops was cancelled in 1968 (6), a high priority problem at the present time is to find ways to restore As-contaminated soils to their optimal level of production (55). Several approaches have been attempted.

Molar P/As Ratio. One approach is to add sufficient phosphate to the system to depress the uptake of arsenate by the plant. Hurd-Karrer (37) showed in solution culture work that this hypothesis was valid and that a molar P/As ratio of at least 5 was needed to protect against As toxicity to wheat. Rumberg et al. (56) also reported that phosphate will improve growth in nutrient solutions containing sufficient arsenate to be toxic. The results with soils systems, however, have been less clear, due in part to the difficulty of evaluating "available" P and As. For example, Jacobs and Keeney (35) found that P additions did not influence As toxicity on a silt loam soil. This soil had a high P fixation capacity and available P probably did not increase greatly. However, with a sandy soil, P actually enhanced As toxicity. They hypothesized that, with the sand, phosphate may have displaced sorbed arsenate resulting in the enhanced toxicity. Similar results were obtained by Woolson et al. (26) with a sandy loam soil. When sufficient phosphate was added to maintain an available P/As ratio of about 7, improved yields resulted. This effect was not consistent, however, and at very high levels of added As (1,000 ppm), phosphate did not overcome As toxicity even at a P/As ratio of 10.

Addition of Fe or Al Compounds. Since As is sorbed by Fe and Al components of soils, another obvious approach is to ammend the soil with Fe or Al salts. Large amounts (5 to 10 metric tons/ha) of ferrous sulfate or ferric sulfate have occasionally reduced As toxicity (57, 58, 59). Steevens et al. (13) attempted to alleviate As toxicity on a sandy soil by application of 4 metric tons/ha of ferric sulfate or aluminum sulfate. The Fe treatment had a slight beneficial effect while the Al treatment actually depressed yields further. Both treatments decreased As uptake by potatoes.

Cultural Practices. Deep plowing to dilute the As concentration of the surface soil and expose As to more sites for fixation would seem to be one of the most economical methods of decreasing toxicity. This approach was suggested by McLean et al. (53) and Albert (31). Vincent (27) also suggested growing tolerant cover crops such as rye or Sudan grass. When plowed under, these crops reduced subsequent As toxicity. Apparently As toxicity to fruit trees is involved with induced Zn deficiency, as foliar application of zinc sulfate or Zn chelates will overcome As toxicity to peach trees (42, 60).

Leaching. Woolson et al. (26) pointed out that in soils in which added phosphate desorbs arsenate, deliberate leaching of the soil after addition of phosphate may be a viable approach to removing As from the root zone. This would seem feasible since the limited data (13, 26, 35) indicates that phosphate most likely will desorb arsenate on sandy soils, which also are easiest to leach since they are very porous.

Literature Cited

1. Kunze, G. W. *In* "Pesticides and their Effects on Soils and Water," pp. 53-54, ASA Spec. Pub. No. 8, Soil Sci. Amer, Inc., Madison, Wi., 1966.

2. Mellanby, Kenneth. "Pesticides and Pollution," pp. 107-108, Collins Press, London, 1967.

3. Crafts, A. S. and R. S. Rosenfels. Hilgardia (1939) 12, 177-200.

4. Mackenthun, Kenneth M. Proc. North Central Weed Control Conf., (1960) 17, 30-31.

5. Cunningham, C. E., T. G. Eastman, M. Goven. Am. Pot. J. (1952) 29, 8-16.

6. U.S.D.A.-A.R.S. Pesticides Regulation Division. Federal Register, May 11, 1968 p. 4.

7. Olson, Oscar E., Lewis L. Sisson, and Alvin L. Moxon. Soil Sci. (1940) 50, 115-118.

8. Williams, K. T. and R. R. Whetstone. U.S.D.A. Tech. Bull. 732, 1940.

9. Fleischer, M. "Cycling and Control of Metals: Proceedings of an Environmental Resources Conference," National Environmental Research Center, Cincinnati, 1973.

10. Wagner, Sheldon L. "Heavy Metals in the Environment," Oregon State University Water Resources Research Institute, Oregon State Univ., Corvallis, Oregon, 1973.

11. Bowen, H. J. M. "Trace Elements in Biochemistry," Academic Press, New York, 1966.

12. Woolson, E. A., J. H. Axley, and P. C. Kearney. Soil Sci. Soc. Amer. Proc. (1971) 35, 938-943.

13. Steevens, D. R., L. M. Walsh, and D. R. Keeney. J. Environ. Quality (1972) 3, 301-303.

14. Deuel, Lloyd E. and Allen R. Swoboda. Soil Sci. Soc. Amer. Proc. (1972) 36, 276-278.

15. Foreback, C. C. "Some Studies on the Detection and Determination of Mercury, Arsenic and Antimony in Gas Discharges," Ph.D. Thesis, Univ. South Florida, 1973.

16. Wood, J. M. Science (1974) 183, 1049-1052.

17. Turner, A. W. Nature (1949) 164, 76-77.

18. Reed, J. E. and M. B. Sturgis. J. Amer. Soc. Agron (1936) 28, 432-436.

19. Quastel, J. H. and P. G. Scholefield. Soil Sci. (1953) 75, 279-285.

20. Black, C. A. *In* T. L. Willrich and G. E. Smith (ed.) "Agricultural Practices and Water Quality," 72-93, Iowa State Univ. Press, Ames, 1970.

21. Johnson, L. R. and A. E. Hiltbold. Soil Sci. Soc. Amer. Proc. (1969) 33, 279-282.

22. Woolson, E. A. "The Chemistry and Toxicity of Arsenic in Soil," Ph.D. Thesis, Univ. Maryland, College Park, Maryland, 1969.

23. Peterson, G. W. and R. B. Corey. Soil Sci. Soc. Amer. Proc. (1966) 30, 563-565.

24. Williams, J. D. H., J. K. Syers, T. W. Walker. Soil Sci. Soc. Amer. Proc. (1967) 31, 736-739.

25. Jacobs, L. W., J. K. Syers, and D. R. Keeney. Soil Sci. Soc. Amer. Proc. (1970) 34, 750-754.

26. Woolson, E. A., J. H. Axley, and P. C. Kearney. Soil Sci. Soc. Amer. Proc. (1973) 37, 254-259.

27. Vincent, C. L. Wash. Agr. Exp. Sta. Bull 437, 1944.

28. Bishop, R. F., and D. Chisholm. Can. J. Soil Sci. (1962) 42, 77-80.

29. Deuel, Lloyd E. and Allen R. Swoboda. J. Environ. Quality (1972) 1, 317-320.

30. Small, H. G., Jr., and C. B. McCants. Agron. J. (1962) 54, 129-133.

31. Albert, W. B. S. Carolina Agr. Exp. Sta. 45th Annu. Rep., 1932.

32. Woolson, E. A. Weed Sci. (1973) 21, 524-527.

33. Tammes, P. M. and M. M. de Lint. Neth. J. Agric. Sci. (1969)

17, 128-132.

34. Benson, N. R. Soil Sci. (1953) 76, 215-244.

35. Jacobs, L. W. and D. R. Keeney. Soil Sci. Plant Analysis (1970) 1, 85-93.

36. Deb, D. L. and N. P. Datta. Plant Soil (1967) 26, 432-444.

37. Hurd-Karrer, A. M. Plant Physiol. (1939) 14, 9-29.

38. Misra, S. G. and R. C. Tiwari. Indian J. Appl. Chem. (1963) 26, 117-121.

39. Vincent, C. L. Proc. 35th Ann. Meeting Washington State Horticultural Assoc. (1939) 82-84.

40. Jacobs, L. W., D. R. Keeney, and L. M. Walsh. Agron J. (1970) 62, 588-591.

41. Arnott, J. T. and A. L. Leaf. Weeds (1967) 15, 121-124.

42. Thompson, A. H. and L. P. Batjer. Soil Sci. (1948) 69, 281-290.

43. Liebig, G. F. *In* H. D. Chapman (ed.) "Diagnostic Criteria for Plants and Soils," pp. 12-23, Univ. Calif., Riverside, 1966.

44. Vandecaveye, S. C., G. M. Horner, and C. M. Keaton. Soil Sci. (1936) 42, 203-215.

45. Cooper, H. P. et al. S. Carolina Agr. Exp. Sta. 45th Ann. Rep. (1931) 23-27.

46. Woolson, E. A., J. H. Axley, and P. C. Kearney. Soil Sci. Soc. Amer. Proc. (1971) 35, 101-105.

47. Liebig, G. R. Jr., G. R. Bradford, and A. P. Vanselow. Soil Sci. (1959) 88, 342-348.

48. Greaves, J. E. Soil Sci. (1934) 38, 355-362.

49. Anastasia, F. B. and W. J. Kender. J. Environ. Qual. (1973) 3, 335-337.

50. Jones, J. S. and M. B. Hatch. Soil Sci. (1945) 60, 277-288.

51. Grimmett, R. E. R. New Zealand J. Agr. (1939) 58, 383-391.

52. Fleming, W. F., F. E. Baker, and L. Koblitsky. J. Econ. Ent. (1943) 36, 231-233.

53. McLean, H. C., A. L. Weber and J. S. Joffe. J. Econ. Ent. (1944) 37, 315-316.

54. Steevens, D. R., L. M. Walsh, and D. R. Keeney. Pesticides Monitoring Journal (1972) 6, 89-90.

55. Allaway, W. H. *In* A. G. Norman (ed.) "Advances in Agronomy," (1968) 20, 235-274.

56. Rumberg, C. G., R. E. Engel, and W. F. Meggitt. Agron. J. (1960) 52, 452-453.

57. Kardos, L. T., S. C. Vandecaveye, and N. Benson. Wash. Agr. Exp. Sta., Bull. 410 (1941).

58. Overly, F. L. Wash. Agr. Exp. Sta. Bull 514, 1950, p. 14.

59. Vandecaveye, S. C., C. M. Keaton, and L. T. Kardos. Proc. Wash. State Hort. Assoc. (1938) 34, 150-158.

60. Batjer, L. P. and N. R. Benson. Proc. Amer. Soc. Hort. Sci. (1958) 72, 74-78.

61. Headden, W. P. Colo. Agr. Exp. Sta. Bull. 131, 1908.

62. Miles, J. R. W. J. Food Agr. Chem. (1968) 16, 620-622.

4

Behavior of Organoarsenicals in Plants and Soils

A. E. HILTBOLD

Agronomy and Soils Department, Agricultural Experiment Station, Auburn University, Auburn, Ala. 36830

During the past 20 years the organic arsenical herbicides have come into extensive use. The more commonly used compounds are sodium or ammonium salts of MAA[1/] (methanearsonic acid), MSMA (monosodium methanearsonate), DSMA (disodium methanearsonate), and MAMA (monoammonium methanearsonate). The methanearsonates differ from inorganic orthoarsenate in having a methyl substitution of one of the hydroxyl groups linked to the arsenic atom. Replacement of another hydroxyl by a second methyl group produces dimethylarsinic or cacodylic acid (hydroxydimethylarsine oxide).

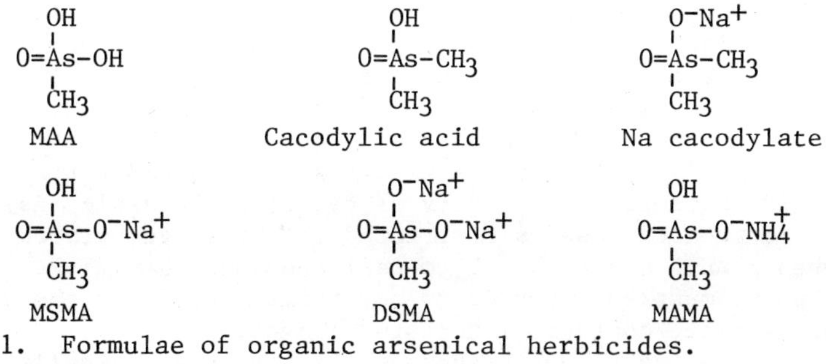

Figure 1. Formulae of organic arsenical herbicides.

Methanearsonic acid is a dibasic acid with pKa values of 3.61 and 8.24 at 18C for dissociation of the first and second OH groups (1). The relative proportion of ionic and molecular forms of methanearsonate is determined by solution pH; Figure 2. At pH 5.93 essentially all of the methanearsonate exists as the univalent ion, or in terms of herbicide formulation as MSMA. The undissociated acid (MAA) increases in solution as pH falls below 5.93 and the proportion of univalent ion decreases. Solutions of MAA are acid, about pH 2. With increasing pH from 5.93 the divalent ion (DSMA) increases in proportion. DSMA solutions are aklaline, about pH 10.5.

1/Abbreviations used in the text are the common names of these herbicides accepted by the Weed Science Society of America.

Cacodylic acid has a single acidic OH; pK_a=6.19 at 25C (1). The proportion of ionic form increases from near zero at pH 4.0 to essentially 100% at pH 8.5, Figure 3. Cacodylic acid solutions are mildly acidic while Na cacodylate solutions are mildly alkaline. While various acid or salt formulations of these organoarsenicals may be applied, their chemical form in soil solution is determined by the existing pH. In most agricultural soils with pH values in the range pH 5.0 to 7.0 the univalent ion of methanearsonate predominates almost to the exclusion of the undissociated or divalent forms. Both cacodylate ion and acid may occur in soils, with the acid form predominant at pH<6.2 and the ionic form predominant at pH>6.2.

Herbicidal Properties and Use Patterns

MSMA, DSMA, and MAMA are available in liquid form. DSMA is marketed also as a white, crystalline powder. Cacodylic acid may be obtained in crystalline form or in mixture with Na cacodylate as a liquid. All of these products are highly soluble in water. They are applied with surfactant as sprays to the foliage of weeds. Theyhave no preemergence activity at rates used for weed control.

Methanearsonates are used most extensively for postemergence control of annual grass weeds in cotton (*Gossypium hirsutum* L.). Activity of the methanearsonates on johnsongrass [(*Sorghum halepense* (L.) Pers.)] purple nutsedge (*Cyperus rotundus* L.), and yellow nutsedge (*Cyperus esculentus* L.) makes these herbicides an important part of the weed control program in cotton. Substantial control of these weeds is provided by the methanearsonates (2), but repeated applications are usually required. MSMA is registered for use in cotton at 2.2 to 2.8 kg/ha and DSMA at 2.2 to 3.4 kg/ha. In order to avoid injury of cotton and arsenic (As) residues in cottonseed, use is restricted to the stage between the time when plants are about 10 cm tall and when they first bloom (3). Sprays must be directed to cover small weeds in the drill row but minimize contact with cotton leaves.

Selective control of crabgrass (*Digitaria sanguinalis* (L.) Scop. and *D. ischaemum* (Schreb.) Muhl.) and dallisgrass (*Paspalum dilatatum* Poir.) in established turf is provided by the methanearsonates with minimal injury of most turfgrass species. Nutsedge is more persistent but may be effectively controlled with repeated applications of 4 kg/ha (4). Methanearsonates are also widely used in non-crop areas for control of johnsongrass and various weeds on rights-of-way, along ditch banks, etc.

Cacodylic acid and Na cacodylate are non-selective, foliar contact-type herbicides (5). Use is restricted to non-crop areas where they are useful in controlling many herbaceous and woody species. Rates required for effective control range from 3 to 5 kg/ha and may be repeated as necessary. Tree injection with cacodylic acid provides selective elimination of undesirable species in established stands. Cacodylic acid is also used in turfgrass

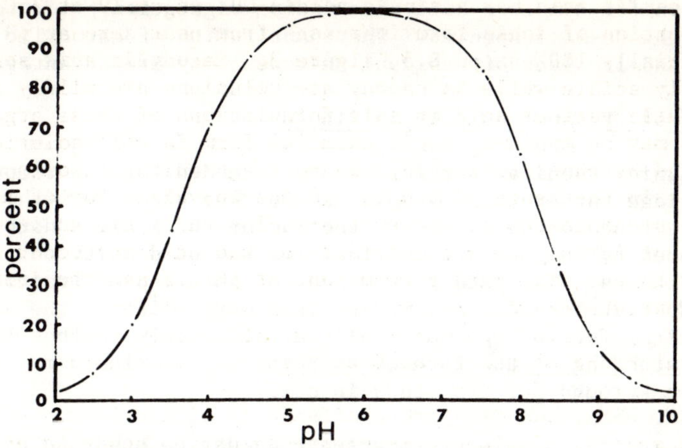

Figure 2. Per cent total methanearsonate as the univalent ion $CH_3AsO(OH)O^-$

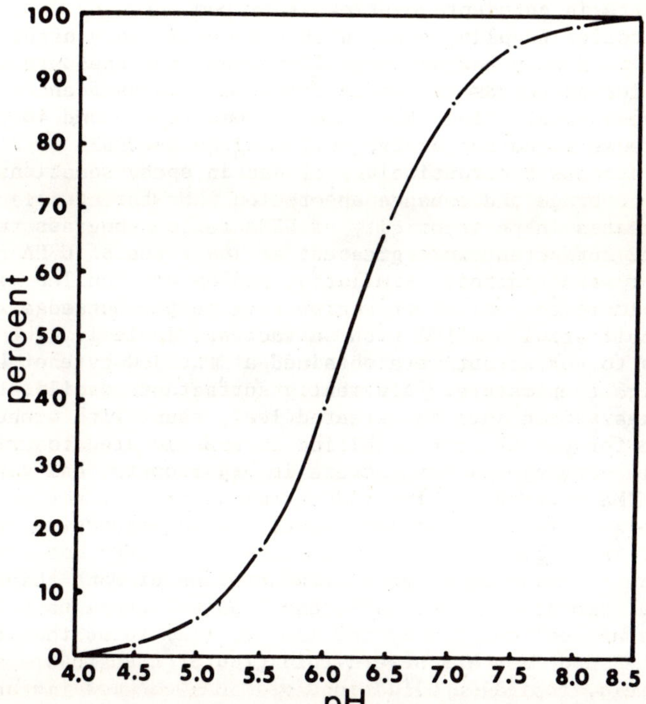

Figure 3. Per cent total cacodylate as the univalent ion $(CH_3)_2AsO\ O^-$

renovation for complete kill of existing vegetation prior to reseeding the desired species. The inactivation of cacodylic acid in soil and its lack of residual phytotoxicity are advantages.

Behavior of Organoarsenicals in Plants

Organic arsenical herbicides are intended for interception by weed foliage rather than by soil as would be the case with a preemergence-type herbicide. The fraction of the applied herbicide intercepted by foliage varies widely with weed density and spraying conditions. The extent to which the intercepted fraction is absorbed, translocated, and metabolized may influence the amount and chemical form of the material that ultimately reaches the soil. An understanding of the fate of an herbicide in plants may provide useful background for its study in soil.

Absorption. While organoarsenicals may be absorbed by roots of johnsongrass (6), Coastal bermudagrass [*Cynodon dactylon* (L.) Pers.] (7), and bean (*Phaseolus vulgaris* L.) (8) in solution culture, root uptake from soil is drastically reduced (7). The primary pathway of entry into plants is through the leaves and stems. Duble, Holt, and McBee (7) compared uptake of As by Coastal bermudagrass from several sources of application of DSMA, Figure 4. Application to foliage was at 4.5 kg/ha, to soil at 17.9 kg/ha, and to roots in nutrient solution at 12 ppm.

Absorption usually occurs within several hours after application (9). Keeley and Thullen (10) found that exposure of yellow nutsedge leaves to MSMA and DSMA for as little as 5 and 15 minutes, respectively, controlled this weed. However, 24- and 48-hr exposures were required for control of purple nutsedge.

Surfactants are routinely included in spray solutions to increase coverage and enhance absorption. McWhorter (11) found that increases in phytotoxicity of DSMA to johnsongrass resulting from added surfactant were greatest at low rates of DSMA near the margin for weed control. Similarly, Keeley and Thullen (10) observed improved control of regrowth of purple nutsedge following foliar application of DSMA with surfactant, Table I. The larger responses to surfactant were obtained at the low rate of DSMA and at the high temperature. Apparently surfactants facilitate entry of methanearsonate into the treated leaf; then, with temperature favorable for growth, the herbicide is translocated to critical sites. In cotton, however, severe injury from topical application of MSMA occurred at low (13C) temperature but not at 31C, a more favorable temperature for growth. This injury was aggravated by surfactant (12).

There is evidence of greater absorption of MSMA than of DSMA. When equal amounts of MAA, MSMA, and DSMA were applied to cotyledons of young cotton, Keeley and Thullen (12) found that absorption of MAA and MSMA was severalfold greater than that of DSMA. In the field, topical applications of MSMA were more injurious to

Table I. Regrowth of Purple and Yellow Nutsedge (% Fresh Shoot Weight of the Control) 3 Weeks Following a Foliar Application of DSMA or MSMA to Plants Grown at Various Temperatures (10)

Herbicide	Surfactant	Temperature 13C Purple	13C Yellow	20C Purple	20C Yellow	29C Purple	29C Yellow
		(Fresh weight as % of control)					
1.68 DSMA	none	100	14	66	27	84	0
1.68 DSMA	0.5%	99	0	77	16	12	0
3.36 DSMA	none	89	0	40	12	77	0
3.36 DSMA	0.5%	54	0	21	0	20	0
1.68 MSMA	none	12	0	48	4	31	0
1.68 MSMA	0.5%	1	0	17	5	32	0
3.36 MSMA	none	7	0	6	0	5	0
3.36 MSMA	0.5%	0	0	0	0	1	0

Weed Science

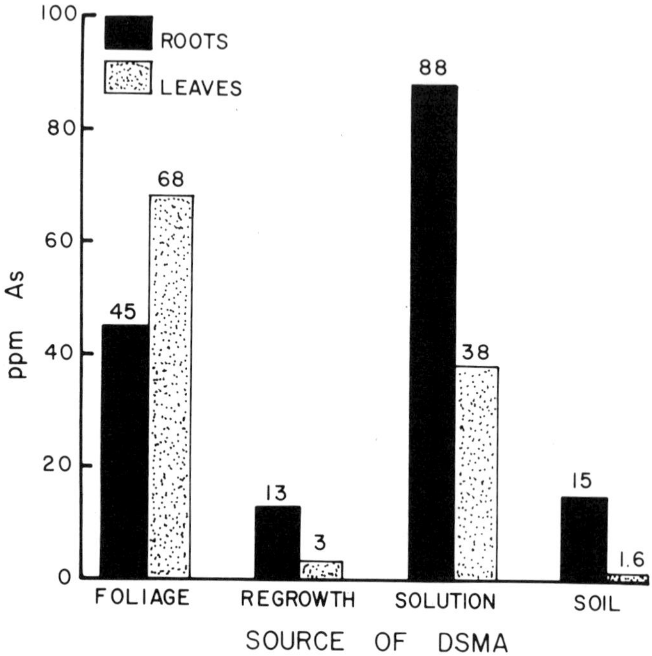

Journal of Agricultural and Food Chemistry

Figure 4. Distribution of As in Coastal bermudagrass 7 days after foliar, nutrient solution, and soil applications of DSMA. Regrowth from a foliar application represented 3 weeks growth after clipping to ground level 7 days after treatment (7).

cotton than equal rates of DSMA, and the difference became more pronounced with higher application rates and later stages of growth (13). Temperature may also influence the relative absorption of MSMA and DSMA. In purple nutsedge growing at 29C, absorption of MSMA and DSMA did not differ, but at 13 and 20C absorption of MSMA exceeded that of DSMA (10). An explanation of the differential absorption of the methanearsonates may lie in the pH of their solutions, their predominant ionic or molecular species, and the permeability of leaf membranes to these species. The evidence suggests that the undissociated acid or univalent ion penetrates more readily than the divalent ion.

Plant species differ in their sensitivity to methanearsonates, probably due in part to differences in permeability of leaf surfaces. Leaves of yellow nutsedge have been found to be more permeable to these herbicides than those of purple nutsedge (10). The tolerance shown by cotton and many turfgrasses for methanearsonates is fundamental to the use of these herbicides for selective weed control. The basis for this tolerance is not yet clear. Tolerance among plant species is relative, however, as evidenced by severe injury of cotton from foliar applications at bloom stage (3).

Translocation. Herbicide movement out of the treated tissue and into more critical sites of activity is essential if the herbicide is to function more than as a desiccant. The superiority of methanearsonates over contact type herbicides such as Na arsenite is due in part to their more extensive translocation (8, 9). Compared to amounts that may be present in treated tissue, however, relatively small amounts are translocated to other plant parts. In soybean [Glycine max (L.) Merr.] and crabgrass, Rumberg, Engel, and Meggitt (9) found that less than 2% of the DSMA recovered 6 hr after application to the leaves was translocated into other plant parts. Sachs and Michael (8) reported that 6% of the MSMA absorbed in treated leaves of bean was translocated during a 3-day period. In purple nutsedge, less than 15% of the applied DSMA was moved out of the treated shoots in 5 days (14). In Coastal bermudagrass, 24% of the applied DSMA was translocated from treated leaves to other plant parts in a 5-day period (7). Difficulty of obtaining adequate translocation of methanearsonates to johnsongrass rhizomes (11) and nutsedge tubers (14, 15) makes repeated applications necessary.

Increasing temperature within the range favorable for growth generally accelerates translocation of methanearsonate. Rumberg et al. (9) reported more rapid translocation of DSMA at 30C than at 15C in soybean and crabgrass. The increased translocation was also associated with increased phytotoxicity of DSMA to crabgrass. Similarly, movement of methanearsonate from treated nutsedge plants into daughter plants was found to be greater at 29C than at 13C (10). In cotton, contact injury of cotyledons by MSMA may preclude translocation to the leaves and terminal bud, sparing the plant severe injury. At low temperature (13C), however, contact injury was negligible and MSMA translocation exceeded that at 31C, resulting in stunting or death of the plant (12).

Both acropetal and basipetal translocation moves methanearsonate out of treated leaves of johnsongrass (6), nutsedge (10, 14), bean (8), Coastal bermudagrass (7), soybean and crabgrass (9). Sckerl and Frans (6) observed very little basipetal movement of methanearsonate in cotton, although foliar and stem application moved acropetally. Basipetal movement of methanearsonate to rhizomes and tubers is critical to the control of hardy perennials such as johnsongrass and nutsedge. Methanearsonate applied to nutsedge shoots is translocated to newly developing rhizomes, shoots, and tubers as indicated by phytotoxic reaction and As determination (14, 16). Accumulation of As in these active growth sites suggests that meristematic tissues are the major points of herbicidal activity.

Metabolism. There is no evidence of substantial degradation of rupture of the C-As bond of organoarsenicals in plants. MSMA-derived CO_2 in the respiratory CO_2 of treated Coastal bermudagrass and purple nutsedge leaves amounts to a small fraction of the herbicide in the tissue (7, 14). Cacodylic acid is apparently a very stable compound in bean plants; chromatographic analysis showed little if any As associated with any fraction other than cacodylic acid itself (8). Extracts of MSMA-treated bean, however, revealed two As-containing fractions: parent MSMA and an unidentified complex of MSMA with some plant component (8). Similar complexes have been observed in johnsongrass after treatment with MAA (6) and Coastal bermudagrass after application of DSMA (7). Radioassay and As analysis of the DSMA complex verified the persistence of the C-As bond (7). The mechanism of phytotoxicity of the organoarsenicals is not known. Critical levels of As required for control of regenerative parts such as nutsedge tubers have not been established (16).

Behavior of Organoarsenicals in Soil

Much of the spray that is not intercepted by foliage is deposited directly on the soil. In addition, much of that which is intercepted by weeds reaches the soil indirectly when rain washes leaf surfaces or killed vegetation decays. While methanearsonate or cacodylate may be applied in acid or salt form, once in the soil the predominant ionic or molecular form is determined by soil solution pH. In most soils (pH 5 to 7) the univalent ion of methanearsonate is the dominant form. Cacodylate exists in approximately equal concentrations of undissociated acid and cacodylate ion in moderately acid soils.

Adsorption. Much of the reduction in activity of organoarsenicals upon reaching the soil is attributable to adsorption by soil colloids. When solutions of DSMA, Na cacodylate, arsenate, and phosphate were equilibrated with 16 soils, Wauchope (17) found that adsorption by individual soils increased in the order: phosphate $<$ Na cacodylate $<$ DSMA $\simeq$ arsenate. Adsorption of all four chemicals was correlated with soil clay content ($r = 0.81$ to 0.96). A similar relationship of DSMA adsorption and soil texture was observed by Dickens and Hiltbold (18) in four surface soils. Subsoils have been found to be much more adsorptive than surface soils for MSMA, probably as a result of their greater contents of clay and iron oxides (19). When several clay minerals of equal particle size were compared as to their adsorption of DSMA, kaolinite was considerably more effective than vermiculite over a range of solution concentrations (18). The greater adsorptive capacity of kaolinite and limonite relative to the 2:1 type clays vermiculite and montmorillonite (Table II) was attributed to the affinity of methanearsonate for mineral surfaces with exposed hydroxyl groups.

The amount of methanearsonate adsorbed by clay increases with the concentration of the equilibrium solution (18). Vaiden clay adsorbed DSMA equivalent to 81 μg As per g soil in equilibrium with 3 ppm As in solution. Results cited by Wauchope (17) indicate adsorption of 2270 μg As per g of clay soil in equilibrium with DSMA solution containing 75 ppm As. These results show great adsorptive capacity in soils, yet, from the standpoint of potential injury to plants, the primary concern must be with the concentration in equilibrium solution.

While initial adsorption of arsenicals is rapid (17), long-term changes result in redistribution of As into less soluble forms (20). During an 8-week period after application of cacodylic acid to three soils, Woolson and Kearney (20) observed declining amounts of water-soluble cacodylic acid and increasing amounts in a less soluble fraction associated with aluminum. Inorganic arsenate in soil has been shown by Jacobs, Syers, and Keeney (21) to become progressively less extractable, with as much as 1 to 6 months required to reach equilibrium. Retention of arsenate was correlated with clay and free iron oxides in 24 soils. Removal of iron oxides and alumina by prior treatment markedly reduced the adsorptive capacity of soil for arsenate (21).

Plant growth in soils containing added organoarsenicals provides another view of the progressive inactivation. Schweizer (22) added DSMA to a silt loam soil at rates extending from 2.5 to 480 ppm. Cotton planted immediately after application was injured at DSMA rates of 80 ppm or greater. In replantings 16 weeks after application, rates of 120 ppm or more were required for injury. Continued amelioration of toxicity was evident in cotton planted 24 and 32 weeks after application.

Addition of phosphate to DSMA-containing soil aggravated the injury of cotton (22). Apparently phosphate displaces methanearsonate from adsorption sites, increasing its activity in the soil solution at the levels of phosphorus applied.

Table II. Adsorption of DSMA from 25 ml of 5.0 ppm As Solutions by 1-g Samples of Augusta Soil Separates and by Certain Reference Materials (18)

Material	Equilibrium solution ppm As	DSMA adsorbed μg As per g	DSMA adsorbed % total applied
Augusta silt loam			
Sand	4.40	15.0	12
Silt	4.43	14.3	11
Clay	0	125.0	100
Limonite	0	125.0	100
Montmorillonite	4.39	15.3	12
Kaolinite	1.20	95.1	76
Vermiculite	3.71	32.3	26

Weeds

Leaching. Packed columns of soil have been used in leaching studies in the laboratory. Ehman (23) reported results from applications of DSMA and cacodylic acid to soil columns 30 cm in depth. Passage of 152 cm of water through the columns removed 9% of the arsenical from a sand soil and 6% from a clay soil. Dickens and Hiltbold (18) applied DSMA to the surfaces of a loamy sand and a clay loam in 23-cm columns at 112 kg/ha on an area basis. Approximately 52% of the applied DSMA was leached from the loamy sand during passage of 76 cm of water. None was leached from the clay loam. Analysis of the clay loam columns revealed nearly half of the applied DSMA in the 0 to 2.5 cm depth with none deeper than 15 cm (18).

Variation of soil pH in the range pH 5.5 to 6.5 did not affect the leaching of methanearsonate in Norfolk loamy sand (18). This is not surprising, considering the dominance of the univalent form of methanearsonate over this pH range. In order for as much as 20% of the applied methanearsonate to exist as undissociated acid or as divalent ion the soil pH would have to be 4.2 or 7.6, respectively. Therefore, pH changes in the range encountered in agricultural soils should not appreciably influence the chemical form of methanearsonate. Soil pH influences the activity of aluminum in clay and hydrous oxides of the soil colloidal system, but apparently the adsorptive capacity of soil is determined more by the amount of these colloids than by pH effects on their activity.

Leaching of arsenic applied as MSMA, DSMA, and MAMA in the field was reported by Johnson and Hiltbold (24). These herbicides were applied repeatedly over a 4-year period for control of nutsedge in turf on a sandy loam soil. At the end of this period the soil was analyzed for As. Greatest concentrations of As were found in the upper 5 cm and decreasing concentrations with depth to 30 cm, Figure 5. The soil was not stirred during the application period; thus, downward movement was attributed to leaching. There were no differences among methanearsonates as to their rate of leaching. Rate of application affected the quantity of As recovered at the several depths but did not appear to influence the rate of downward movement. Results in this soil indicated a slow but perceptible movement of As into subsurface horizons to at least 30 cm depth (24). Further field experiments of Hiltbold, Hajek, and Buchanan (19) showed less mobility of As in soil profiles. After 6 years of MSMA applications to soils growing cotton, As distribution in the profiles showed little if any evidence of leaching below the plow depth, 23 cm. The chemical form of As in these soils at the time of their analysis was not determined.

Metabolism. In contrast to the stability of the C-As bond observed in methanearsonates in plants, microbiological oxidation occurs in soils. Dickens and Hiltbold (18) found methanearsonate-derived CO_2 evolved from 4 soils in amounts ranging from 0.7 to

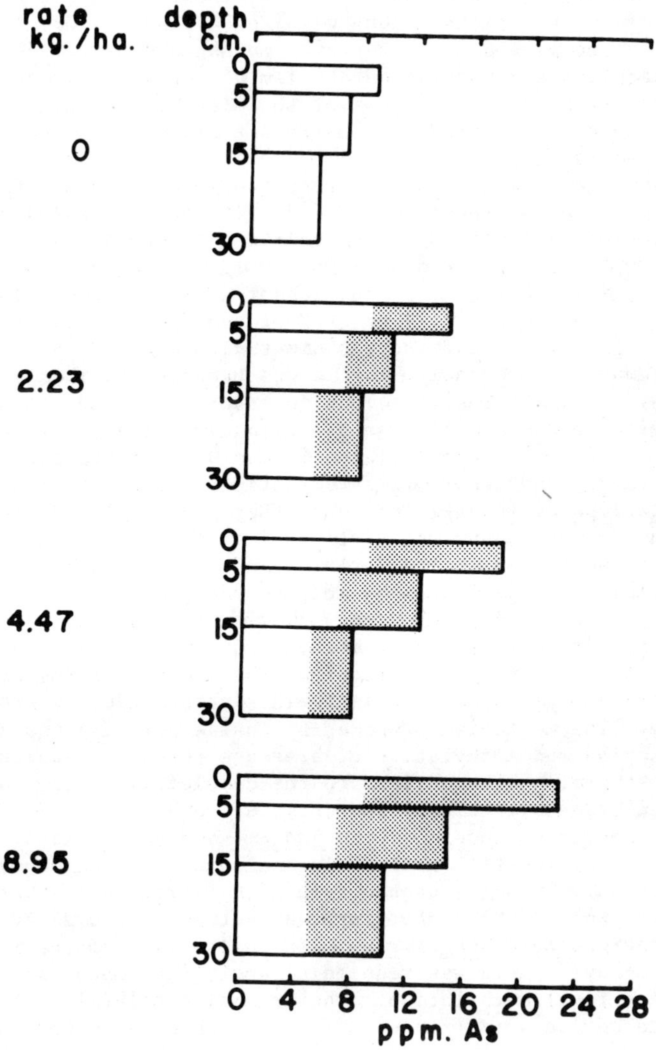

Figure 5. Distribution of As in soil after applying methanearsonates at three rates for 4 years. Total As applied equivalent to 16, 32, and 64 kg As/ha. Shaded portions represent recovery of applied As, averaged across MSMA, DSMA, and MAMA treatments (24).

5.5% of the added MAA during a 30-day incubation. Von Endt, Kearney, and Kaufman (25) reported oxidation of the methyl carbon of MSMA in 4 soils amounting to 1.7 to 10% during a 3-week period. Methanearsonate oxidation in these soils was proportional to their organic matter contents (25) or to their total CO_2 evolution (18). These observations, along with the dependency of microbial isolates on energy sources other than methanearsonate (25), indicate a predominantly passive metabolism of methanearsonate in soil, with little response on the part of the microbial population. Arsenate has been identified as the product of methanearsonate oxidation in soils (25).

Woolson and Kearney (20) reported two loss processes for cacodylic acid in soils. Under aerobic conditions, approximately 41% of the applied cacodylic acid was oxidized to CO_2 and arsenate, while 35% was lost from the soil as a volatile organoarsenical compound during a 24-week period. Under anaerobic conditions, degradation via CO_2 was nil, yet 61% of the applied cacodylic acid was volatilized, possibly as dimethyl arsine.

A gaseous loss mechanism for As was proposed by Thom and Raper (26) in 1932. They observed in laboratory studies the microbiological reduction of arsenate with evolution of arsenical gases. Active fungi were isolated from arsenic-toxic field soils. Challenger (27) and coworkers later defined the methylation of As by *Scopulariopsis*, *Aspergillus*, and other fungi. Their studies showed the reduction and methylation of arsenicals including As_2O_3, Na_2HAsO_4, DSMA, Na cacodylate, and Na propylarsonate. Trimethylarsine [$(CH_3)_3As$] was the most common gaseous product. In all cases, As was reduced to the trivalent state and all substituent oxygen atoms replaced by methyl groups. With methanearsonate, cacodylate, and propylarsonate, the evolved arsine contained the original alkyl intact: trimethylarsine and dimethyl-n-propylarsine. The mechanism proposed by Challenger (27) was a stepwise reduction and methylation of arsenate through arsenite, methanearsonate, and cacodylate to the completely reduced and methylated trimethylarsine. Recently, Cox and Alexander (28) isolated species of *Candida*, *Gliocladium*, and *Penicillium* producing trimethylarsine from arsenicals including MAA and cacodylic acid. The only evidence of bacteria methylating As is that of McBride and Wolfe (29), reporting the synthesis of dimethylarsine by the anaerobe *Methanobacterium*. In this mechanism the intermediate cacodylic acid was reduced directly to dimethylarsine without the final methylation. The extensive volatilization of As from cacodylic acid-treated soils (20) when incubated anaerobically would necessarily implicate bacteria such as *Methanobacterium*. Under aerobic conditions, fungi appear to be the active microorganisms.

Persistence. Loss of phytotoxicity with time after application of organoarsenicals to soil may be viewed as a measure of persistence. Results of Schweizer (22) and Ehman (23) indicate extensive inactivation and diminishing effects on subsequent crops. Yields of a variety of field crops were unaffected by residues of methanearsonate equivalent to 68 kg As/ha applied during the previous 4 years (24). Similarly, yields of cotton were unaffected by 6 annual preplant applications of MSMA at 10, 20, and 40 kg/ha (19). On the other hand, Singh and Campbell (30) found that both DSMA and MAMA, when applied to turf at rates of 5.0 and 4.5 kg/ha, respectively, twice annually for 2 years, persisted in the upper 5 cm of a silt loam soil in amounts sufficient to injure oats (*Avena sativa* L.) and soybeans, as shown by bioassay of soil 9 months after the last application.

Estimates of the rate of methanearsonate oxidation to CO_2 and inorganic arsenate are provided by soil incubations with ^{14}C-labeled methanearsonate (18, 25). In light texture soils of low organic matter content, 2% per month appears to be a reasonable estimate of oxidation. In clay soils with more organic matter and larger microbial populations, about 10% per month may be oxidized. If these rates remain constant, as would be the case in a first-order reaction, the amount of residual methanearsonate at any time after application may be calculated. A first-order function appears to be the simplest and most realistic representation of methanearsonate metabolism in soil. This is supported by the observation of equal percentage losses among widely differing rates of application (20, 25). Moreover, the general lack of microbial response to added methanearsonate (18, 25) suggests a passive degradation dependent primarily upon concentration. This contrasts with the degradation of the phenoxyalkanoic acid herbicides, for example. With these, microbiological adaptation occurs, resulting in rapidly accelerated decomposition, elimination of the lag period, and development of the enrichment effect (31).

The first-order rate law may be expressed as follows:

$$k = \frac{2.303}{t} \log \frac{C_o}{C} \quad \text{and} \quad t_{1/2} = \frac{0.693}{k}$$

where:
k = rate constant
t = time
C_o = initial concentration
C = concentration at time t
$t_{1/2}$ = half-life of decomposition

Substituting and solving for 2 and 10% decomposition per month yields values of 0.7847 and 0.2827, respectively, for concentrations remaining 1 year after application of unit concentration of methanearsonate. Half-lives corresponding to 2 and 10% per month loss rates are 34.31 and 6.58 months, respectively. The residue of N annual additions is given by the equation (32):

$$\frac{r}{C_o} = 1 + f_1 + f_1^{\ 2} + f_1^3 + \text{----------} f_1^{n-1}$$

$$\left(\frac{r}{C_o}\right)_\infty = \frac{1}{1-f_1}$$

where: f_1 = fraction left after one year
r = accumulated residue, immediately after addition of annual increment.

The limit of the maximum amount of residue $\left(\frac{r}{C_o}\right)_\infty$ is given by the expression $\frac{1}{1-f_1}$. The residue pattern for annual additions of 1 ppm MSMA-As is given in Figure 6, assuming first-order decomposition to inorganic arsenate at 10% per month (above) and 2% per month (below). No losses of As from the soil are considered. Annual application of 1 ppm MSMA-As in the plow layer is approximately that provided by two postemergence sprays of MSMA in cotton at 2.5 kg/ha each. At the higher decomposition rate the maximum amount of residual MSMA is 1.39 ppm and is approached after several applications. Residual MSMA in this situation remains at very low levels and the principal product accumulating is arsenate. At the low rate of decomposition the maximum amount of residual MSMA is 4.64 ppm and is approached after 30 annual additions. At this time the soil residues would be 15% as MSMA and 85% as arsenate.

The foregoing estimates of residue accumulation do not take into account losses of As from the soil. Field experiments (19, 24) indicate loss of As from soil during periods of repeated application of methanearsonate. In situations where loss of As in soil erosion and crop removal were nil, approximately 40 to 60% of the As applied as methanearsonate was unaccounted for in total As determinations of the soil. These results suggest a gaseous loss process of considerable important with regard to residue accumulation. Preferential methylation and volatilization of methanearsonate would further reduce residual concentrations and alter the proportion of methanearsonate and arsenate.

Uptake of As by Crops. The literature dealing with inorganic As in soils and crops shows that plants generally absorb and translocate relatively small amounts of As. Little information is available on the appearance of As residues in crops resulting from root absorption of methanearsonate or methanearsonate-derived As. Ehman (23) reported significant levels of As in several field crops planted immediately after application of DSMA to soil, but rates of 35 kg/ha or more were required. Johnson and Hiltbold (24) reported As residues in field crops grown on plots receiving methanearsonate applications during the previous 4 years. Cottonseed and soybeans contained higher levels of As than corn (*Zea mays* L.) grain or vegetative material of sorghum-sudan hybrid (*Sorghum vulgare* Pers. X *S. sudanense* (Piper) Stapf.). Cool-season forages accumulated less As than summer crops.

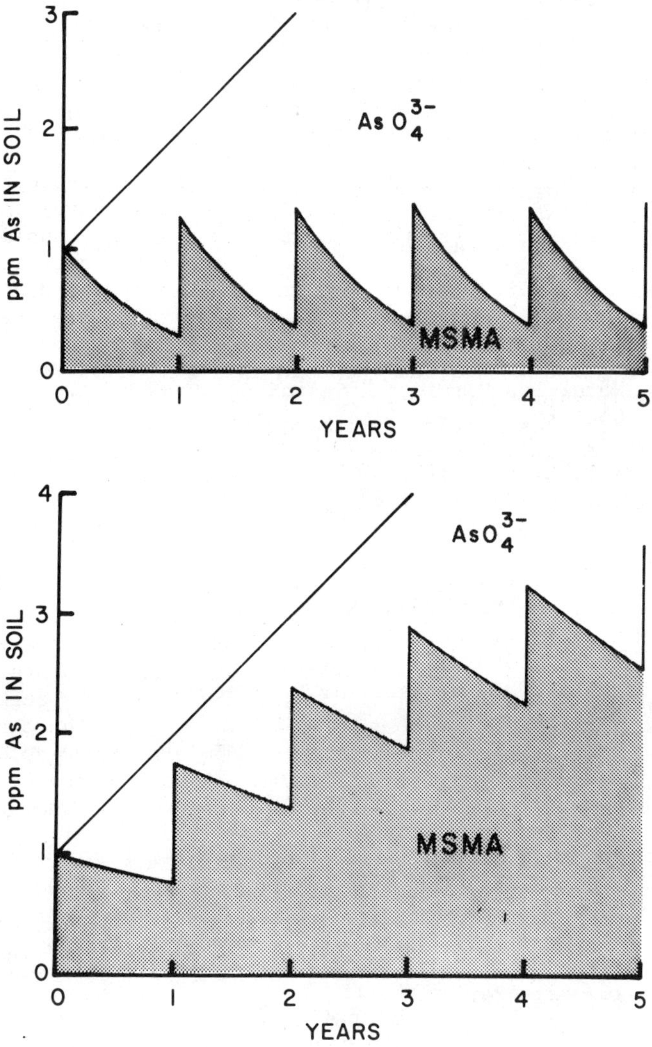

Figure 6. Residue pattern for annual additions of 1 ppm MSMA–As and first-order half-life of 6.58 mo (top) and 34.31 mo (bottom)

Later experiments (19) with soil applications of MSMA at elevated rates preplant to cotton over a 6-year period failed to show As residues in cottonseed grown on 3 soil types.

Sustained used of methanearsonates at rates recommended for weed control does not appear to result in hazardous accumulation of As in harvested crops. Methanearsonates are extensively adsorbed in soil and subject to little movement in leaching. Loss from treated fields may occur where erosion is severe. Volatilization of methylated As appears to be a significant loss process that limits As accumulation in soil and probably accounts for the ubiquitous occurrence of As in soils generally. The organoarsenicals are oxidized microbiologically producing arsenate. While arsenate is subject to methylation and volatile loss, much of the residual As in soil reverts to progressively less soluble forms with aluminum and iron compounds.

Literature Cited

1. Sillen, L. G., Martell, A. E., "Stability Constants of Metal-Ion Complexes" Second Edition, P. 206. Chemical Society, London, 1964.
2. Widiger, R. E., Proc. S. Weed Conf. (1966) 19, 51-56.
3. Baker, R. S., Arle, H. F., Miller, J. H., Holstun, J. T. Jr., Weed Sci. (1969) 17, 37-40.
4. Long, J. A., Allen, W. W., Holt, E. C., Weeds (1962) 10, 285-287.
5. Stevens, G. D., Proc. S. Weed Conf. (1966) 19, 545-549.
6. Sckerl, M. M., Frans, R. E., Weed Sci. (1969) 17, 421-427.
7. Duble, R. L., Holt, E. C., McBee, G. G., J. Agr. Food Chem. (1969) 17, 1247-1250.
8. Sachs, R. M., Michael, J. L., Weed Sci. (1971) 19, 558-564.
9. Rumburg, C. B., Engel, R. E., Meggitt, W. F., Weeds (1960) 8, 582-588.
10. Keeley, P. E., Thullen, R. J., Weed Sci. (1971) 19, 601-606.
11. McWhorter, C. G., Weeds (1966) 14, 191-194.
12. Keeley, P. E., Thullen, R. J., Weed Sci. (1971) 19, 297-300.
13. Arle, H. F., Hamilton, K. C., Weed Sci. (1971) 19,545-547.
14. Duble, R. L., Holt, E. C., McBee, G. G., Weed Sci. (1968) 16, 421-424.
15. Hamilton, K. C., Weed Sci. (1971) 19, 675-677.
16. Holt, E. C., Faubion, J. L., Allen, W. W., McBee, G. G., Weeds (1967) 15, 13-15.
17. Wauchope, R. D., Proc. S. Weed Sci. Soc. (1974) 27, 389.
18. Dickens, Ray, Hiltbold, A. E., Weeds (1967) 15, 299-304.
19. Hiltbold, A. E., Hajek, B. F., Buchanan, G. A., Weed Sci. (1974) 22, 272-275.
20. Woolson, E. A., Kearney, P. C., Environ. Sci. Technol.

(1973) 7, 47-50.
21. Jacobs, L. W., Syers, J. K., Keeney, D. R., Soil Sci. Soc. Amer. Proc. (1970) 34, 750-754.
22. Schweizer, E. E., Weeds (1967) 15, 72-76.
23. Ehman, P. J., Proc. S. Weed Conf. (1965) 18, 685-687.
24. Johnson, L. R., Hiltbold, A. E. Soil Sci. Soc. Amer. Proc. (1969) 33, 279-282.
25. Von Endt, D. W., Kearney, P. C., Kaufman, D. D., J. Agr. Food Chem. (1968) 16,17-20.
26. Thom, C., Raper, K. B., Science (1932) 76, 548-550.
27. Challenger, F., Chem. Rev. (1945) 36, 315-361.
28. Cox, D. P., Alexander M., Bull. Env. Cont. Tox. (1973) 9, 84-88.
29. McBride, B., Wolfe, R., Biochem. (1971) 10, 4312-4317.
30. Singh, R. K. N., Campbell, R. W., Weeds (1965) 13, 170-171.
31. Audus, L. J., Plant and Soil (1951) 3, 170-192.
32. Hamaker, J. W., Advan. Chem. Ser. (1966) 60, 122-131.

5

Arsenicals in Animal Feeds and Wastes

C. C. CALVERT

Biological Waste Management Laboratory, Agricultural Environmental Quality Institute, Agricultural Research Service, Department of Agriculture, Beltsville, Md. 20705

Since arsenic trioxide was first obtained from copper smelting over 4000 years ago, this element has been used for a variety of medicinal purposes (1). The modern era of arsenic use in medicine began with the characterization of arsanilic acid by Erlich and Bertheim in 1907 (2) and was extended more recently by the discovery in 1945 by Morehouse and Mayfield (3) that an organic arsenical, 3-nitro-4-hydroxyphenylarsonic acid commonly called (3-nitro), could be used to control coccidiosis and promote growth in chicks. Since that time, other organic arsenicals such as arsenosobenzene, arsanilic acid, 4-nitrophenylarsonic acid and p-ureidobenzenearsonic acid have been shown to have both therapeutic and growth-promotant properties as feed additives for poultry and swine.

At the present time, all of these compounds, with the exception of arsenosobenzene, are approved by the U.S. Food and Drug Administration for use in poultry and swine feeds but only at levels low enough to preclude residues in edible animal tissue which would be a hazard to human health. The objective of this discussion will be to review briefly the efficacy of these arsenic compounds with respect to animal production, the absorption and excretion of these arsenicals when fed to animals and, finally, the fate of arsenic in animal excreta.

Arsenicals in Swine and Poultry Feeds

Table 1 shows the chemical structure of 4 arsenicals currently approved for use as growth promotants or therapeutic agents, or both, in poultry and swine feeds. The arsenic in all these compounds is pentavalent. Changing the substituents on the ring changes the growth-promoting and anti-parasite effects of the compound. Note that arsanilic acid and 3-nitro are approved for both poultry and swine, whereas 4-nitrophenylarsonic acid and p-ureidobenzenearsonic acid are approved only for turkeys. The use of these compounds in turkeys is limited to feeding for therapeutic purposes and may not be used as growth promotants (4).

Table 1. Chemical structure of arsenicals used in animal feeding

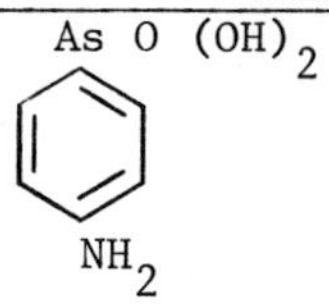 Arsanilic Acid (Poultry & Swine)	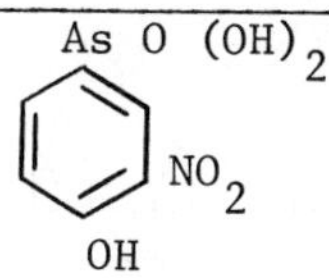 3-nitro-4-hydroxyphenylarsonic Acid (Poultry & Swine)
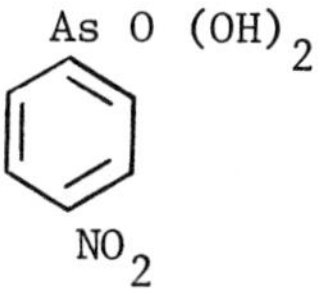 4-nitrophenylarsonic Acid (Turkeys)	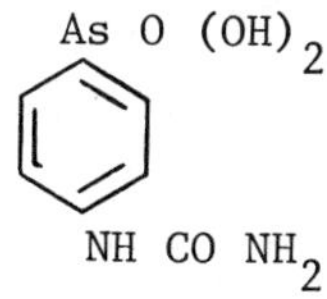 p-ureidobenzenearsonic Acid (Turkeys)

The FDA has established levels at which these compounds may be used in animal feeds, and it has also determined the maximum levels of arsenic that may be present in marketed poultry and swine. These maximum use levels and maximum tissue levels are shown in Table 2. In order that these tissue levels may be met, the user of arsenical-containing feed is required by FDA regulations to withdraw the arsenic-containing feed for at least 5 days before slaughter.

Growth Response from the Use of Arsenicals

Many experiments have expanded on the original demonstration of the efficacy of 3-nitro and other arsenicals in both poultry and swine rations. No attempt will be made in this discussion to cover all of the research that has been conducted with the arsenicals in feed, but a few examples of the types of responses obtained will be presented. A number of reviews on this subject are in the literature and should be consulted for more detailed information (5-9).

Baron (9), at a symposium in London in 1969, presented examples of the type of responses that may be obtained by feeding arsenicals to growing chickens. Results of a study in which 3-nitro was fed in 15 trials to a total of 1148 birds for a 4-week period are shown in Table 3. As can be seen in this table, the average response in weight increase due to arsenic in these studies was 4.1%. The response in feed to gain ratio was 1.6%. The results shown in Table 4 show the average responses from 5 trials involving more than 2500 chickens for an 8-week feeding period. This period represents the time needed to bring a

Table 2. Maximum permissible levels of arsenicals in animal feeds and maximum permissible levels of arsenic in animal tissue.[1/]

Compound	Species	Maximum feed level	Maximum tissue arsenic level
Arsanilic acid	Poultry[2/]	90g/ton(100 mg/kg)	0.5 mg/kg fresh, uncooked muscle 2.0 mg/kg fresh, uncooked by-products
	Swine	90g/ton(100 mg/kg)	0.5 mg/kg fresh muscle and by-products other than kidney & liver 2.0 mg/kg fresh, uncooked kidney & liver
3-nitro-4-hydroxyphenyl-arsonic acid	Poultry[2/]	45g/ton(50 mg/kg)	Same as arsanilic acid
	Swine	68g/ton(75 mg/kg)	Same as arsanilic acid
4-nitrophenyl-arsonic acid	Turkeys	170g/ton(187 mg/kg)	Same as arsanilic acid
p-ureidobenzene-arsonic acid	Turkeys	340g/ton(375 mg/kg)	Same as arsanilic acid

1/ Source: (4).
2/ Broilers, laying hens and turkeys.

chicken to market size. In this test, the increase in weight was not significant, but feed conversion with arsenic supplement was significantly improved - 1.99 g of feed to produce 1 g of live weight gain with the arsenical compared with 2.04 g of feed for each gram of gain with the controls.

Table 3. Summary of tests to show the effect of feeding 3-nitro-4-hydroxyphenylarsonic acid (0.005%) on the weight gain and feed efficiency of broiler chicks during the first 4 weeks of life.[1/]

Treatment	Trials	Birds	Average weight	Average weight response	Average feed conversion	Average feed conversion response
	(No.)	(No.)	(g)	(%)	(g feed/g weight)	(%)
3-Nitro	15	1148	530	4.1	1.70	1.6
Nonmedicated controls	15	1148	510	-	1.73	-

1/ Source: (9).

Table 4. Summary of tests to show the effect of feeding 3-nitro (0.005%) on the growth and feed efficiency of broilers during the first 8 weeks of life.[1/]

Treatment	Trials	Birds	Average weight	Average feed conversion
	(No.)	(No.)	(g)	(g feed/g weight)
3-Nitro	5	2568	1448	1.99
Nonmedicated controls	5	2568	1430	2.04

1/ Source: (9).

Results of an investigation by Hansen (10) on the response of growing swine to 3-nitro are shown in Table 5. The arsenical was fed in this study for the full growing period and resulted in a significant improvement in both average daily gain and feed conversion.

Table 5. Effect of 3-nitro on growth rate and feed conversion of growing sheep.1/

	Average daily gain from weaning to 90.9 kg (kg)	Feed conversion (kg feed/kg weight)
3-nitro (40 mg/kg of feed)	0.79	3.50
Nonmedicated controls	0.75	3.56

1/ Source: (10).

Reporting these results is not intended as an endorsement for 3-nitro as the arsenical of choice in animal feeding. Similar responses have been shown with other arsenicals such as arsanilic acid for chickens, turkeys and growing swine. In the experiments reviewed for this report, no negative responses to arsenical additives were obtained if supplementation did not exceed recommended levels. There were a few instances reported in which no improvements in growth and feed efficency were obtained from the addition of arsenicals to swine and poultry diets, but the high frequency of positive effects from arsenicals gives the producer that small edge that may well be the difference between a profit and loss in a modern swine and poultry feeding enterprise.

Arsenicals are also approved for feeding in combination with other feed additives such as coccidiostats and antibiotics. There have been many experiments reported in the literature testing combinations of the arsenicals with the wide variety of antibiotics and coccidiostats used in poultry and swine diets. Two reviews of these experiments (5,9) should be consulted for more detailed information. In many instances, a combination of arsenical and antibiotic has improved growth that is greater than that obtained with either of the additives fed alone. There is at present no explanation as to why one antibiotic will respond one way when combined with an arsenical and another will respond another way.

Mechanism of Action

A number of theories have been proposed for the mechanism of action of arsenicals in increasing growth in swine and poultry. Peoples (8) has suggested that the arsenical may inhibit those organisms that cause thickening of the gut wall and therefore result in more efficient absorption of nutrients. Another theory is that these compounds act like antibiotics and inhibit harmful bacteria. However, studies of the bacteriostatic value of arsenicals by Frost and Spruth (6) indicate that with certain bacteria, *Escherichia coli* and *Clostridium perfringens*, for example, the arsenicals are much less effective than conventional antibiotics. Another theory, proposed by Pope and Schaible (11), suggests that arsenicals may exert a sparing effect on protein. In this study, egg production in hens fed a 13% protein diet was increased by the addition of arsanilic acid to the diet to a level comparable with the egg production in hens fed a 16.5% protein ration. Russo et al. (12) also found that arsanilic acid reduced urinary nitrogen excretion in swine, and they suggested that arsenicals exert a sparing effect on protein by reducing protein catabolism. Thus, no one mechanism can likely explain the action of an arsenical, and probably a combination of factors produces the observed responses.

Uptake and Depletion of Arsenic from Tissue

Arsenicals in an animal's feed will result in arsenic residues in tissues. Data in Table 6 show that the feeding of arsanilic acid at 0.01% of a chick diet and 3-nitro-4-hydroxyphenylarsonic acid at 0.005% results in low levels of arsenic in liver tissue. Feeding 10 times the approved levels of either of these arsenicals did not result in 10 times the level of arsenic in the tissues. The data suggest that in chicken liver tissue, there is an upper limit for arsenic. These same authors also presented evidence that arsenic in liver and muscle of chicks quickly reached a plateau level and did not increase further when arsanilic acid was fed continuously at 0.005% of the ration for a 9-week period.

Table 6. Arsenic found in livers of chickens fed various arsenicals.[1/]

Compound and feed level	Arsenic (As_2O_3) in- Feed	Fresh liver
	------- ppm	--------
Arsanilic acid:		
0.01%	45.4	1.2
0.1%	455	6.4
3-nitro-4-hydroxyphenylarsonic acid:		
0.005%	18.7	2.4
0.05%	187	7.5
Dodecylamine p-chlorphenylarsonate:		
0.01%	23.3	2.9

1/ Source: (6).

Evidence for the rapidity with which arsenic is taken up and eliminated from chicken tissues is shown in Table 7. These data by Baron (9) show that in the kidney, liver, muscle and skin of chicks fed 0.005% 3-nitro, there is a low-level uptake of arsenic that plateaus after 7 days of feeding and is well below tolerance levels after only 1 day of withdrawal of the arsenical.

Table 7. Summation of arsenic levels (3-nitro, .005% of the diet) in chicken tissues.[1/]

Days									
On Test	1	7	56	70	71	75	80	84	
On Med.	1	7	56	70	-	-	-	-	Non-
Off Med.	-	-	-	-	1	5	10	14	Med.
PPM Arsenic									
Kidney	.93	.76	.52	.64	.22	.10	.09	.08	.05
Liver	1.31	2.43	1.26	1.26	.69	.43	.32	.19	.08
Muscle	.03	.07	.05	.04	.03	.01	.02	.02	.02
Skin	.05	.11	.06	.05	.08	.02	.03	.03	.02

1/ Source: (9).

Feeding arsanilic acid to swine also results in uptake of arsenic in liver, kidney and spleen and, to a lesser extent, in muscle and blood. The method of feeding, and in particular the water intake, may also influence the deposition of tissue arsenic. Vorhies et al. (13) reported that arsenic in liver increased from 1.8 ppm to 3.3 ppm when water intake in swine was reduced. An earlier study by Bridges et al. (14) also showed this effect. None of the arsenicals have been approved by FDA for use in feed for ruminant animals (beef cattle, dairy cows or sheep). However, the current interest in the use of animal wastes, particularly dried poultry excreta, as animal feed supplements has generated interest in the arsenic content of these wastes and in the effect of arsenic on the ruminant. Studies on the uptake of arsenic from arsenicals in sheep have been conducted at our laboratories at Beltsville. In these investigations, arsenic, as arsanilic acid, has been fed to mature wethers at levels up to 273 mg/kg of diet for a period of 28 days. The levels of arsenic found in various tissues is shown in Table 8. All tissues showed increasing levels of arsenic as the dietary arsenic levels increased.

Table 8. Arsenic levels in tissues of wethers fed arsanilic acid for 28 days.

Arsenic fed	Whole blood	Liver	Kidney	Muscle
(mg/kg of diet)		------- mg/kg dry tissue ------		
0.0	<0.01	<0.01	<0.01	<0.01
26.8	0.063	3.1	3.2	0.2
144.4	0.270	26.8	12.2	1.1
273.3	0.536	29.2	23.6	1.2

At the end of the 28-day feeding period, all sheep were placed on an arsenic-free diet, and every 2 days an animal from each treatment group was killed for sample collection. Table 9 shows the results of a 6-day depletion period in liver arsenic. At all levels of arsenical feeding, there was a rapid depletion of arsenic from liver tissue to levels that are below the tolerance levels established by FDA for edible tissue from poultry and swine.

Table 9. Depletion of arsenic in liver of wethers fed arsanilic acid for 28 days.

Arsenic Fed	Withdrawal time (days) 0	2	4	6
(mg/kg of diet)	----- mg/kg D.M.			-----------
0.0	<0.01	<0.01	<0.01	<0.01
26.8	3.1	4.9	2.9	1.9
144.4	26.8	15.4	8.4	3.5
273.3	29.2	27.0	11.4	5.0

As described in this section, arsenic is deposited in animal tissue after low-level feeding, but in all species studied, there is a rapid excretion of this element after withdrawal of the arsenical from the feed.

Metabolism of Arsenic

The metabolism of ingested arsenic from the arsenicals has been investigated by a number of researchers. Moody and Williams (15-17) in a series of experiments showed that chickens excreted arsanilic acid largely unchanged without evidence that it was converted into any other organic arsenic compound or to inorganic arsenic. Four-nitrophenylarsonic acid, however, was partly converted to arsanilic acid and 3-nitro was partly converted to 3-amino-4-hydroxyphenylarsonic acid. There was no indication that any of the 3-nitro was converted to inorganic arsenic.

Frost (18), Hansen et al.(19) and Overby and Frederickson (20) all showed that in poultry and swine, a high percentage of ingested arsenic is excreted and that this excretion is very rapid. In our own work with growing sheep in a 5-day balance trial, about 87% of all ingested arsenic was excreted. Table 10 presents the results of this study. Results from this study also show that about 76% of the 3-nitro in poultry waste fed to sheep was excreted in feces.

Table 10. Arsenic excretion in wethers fed dried broiler manure during a 5 day balance trial.[1/]

Ration	Dried[2/] poultry manure	Arsenic in feed	Total arsenic intake	Total fecal arsenic	Total urinary arsenic	Arsenic excreted
	(%)	(mg/kg DM)	----------	(mg)	----------	(%)
1	0	<0.01	<0.01	<0.01	<0.01	<0.01
2	7	3.42	14.3	8.5	3.4	83.2
3	14	4.95	25.1	18.4	4.4	90.8

1/ Feed intake measured and total feces and urine collected during the last 5 days of a 15 day feeding period.

2/ Contained arsenic from 3-nitro-4-hydroxyphenylarsonic acid in dehydrated broiler manure.

Fate of Excreted Arsenic

There is concern over the extent to which arsenic in animal excreta used as a fertilizer or as a nutrient source in animal feeds presents a hazard to crops or the food chain.

One study conducted by Morrison (21) showed no increase in arsenic in soil, water or forage after poultry litter containing from 15 to 30 ppm arsenic had been applied to land at a rate of 4 to 6 tons per acre per year for 20 years. The rate of arsenic application was approximately 100 g per acre per year, and the amount found in the soil was approximately 1.8 ppm. This level was comparable with that found in soils with no poultry litter added. Volatilization and leaching undoubtedly would account for the disappearance of the 2000 g of arsenic deposited over the 20-year period. Undoubtedly, higher levels of manure will be used, which could result in higher levels of the arsenical residues being deposited on various kinds of soils. However, levels of arsenic applied under these conditions will not likely begin to approach the levels reported when various herbicides and pesticides have been used on crops and soils for long periods of time. To produce a level of arsenic in soil comparable to the 550 ppm found by Williams and Whetstone (22), 2200 tons of poultry litter per acre per year containing 15 to 30 ppm arsenic would have to be applied.

In a recent experiment by Woolson (23), the persistence and chemical distribution of arsanilic acid in Lakeland sandy loam, Hagerstown silty loam and Christiana clay loam were examined. Two application rates (158 and 790 ppm arsanilic acid) and two moisture levels (75 percent of soil capacity and flooding to establish aerobic and anaerobic conditions respectively) were imposed on the three soil types. After a 32-week incubation period less than 10 percent of applied arsanilic acid could be extracted from any soil type. Total arsenic remained constant in the Lakeland and Hagerstown soils and decreased somewhat in the Christiana soil sample. The disappearance of arsanilic acid occurred more rapidly under anaerobic conditions in all soil types. It appears that under the conditions of this experiment arsanilic acid is rapidly degraded to arsenate and an organic fraction and this degradation proceeds more rapidly in anaerobic soils. There seems to be some reduction of arsanilic acid to a volatile organic arsenic compound in Christiana soil.

Processed animal wastes are being considered for use as a feedstuff for certain classes of livestock and poultry; consequently there has been some concern with respect to feed additive residues, including arsenic, in processed animal wastes. The regulations regarding waste recycling will likely specify maximum residue levels that will not be harmful to animals or present a hazard to human health. The final proposal has not been prepared by the FDA at the present time; so we cannot state exactly what will be required with respect to arsenic and other feed additives.

Summary

In summary, arsenicals are used as feed additives to promote growth in poultry and swine as well as to treat specific diseases of these species and appear to have an established role in modern livestock and poultry production. The responses in general are small but may represent the difference between a profit and loss to the animal producer.

Arsenic can be detected at low levels in tissues of animals receiving arsenicals. However, this arsenic is rapidly eliminated when arsenicals are removed from the feed for the required 5-day period. This process appears to be well regulated, and there does not appear to be a hazard to the consumer from arsenic in edible animal tissue.

The arsenicals appear to be largely excreted unchanged in chemical structure with no toxic metabolites produced and with no evidence of the conversion to inorganic forms of arsenic in the animal. Because arsenicals are excreted for the most part in feces, no evidence is available to indicate that the amounts of arsenic that might be incorporated into soils by the disposal of animal wastes would constitute any problem with regard to plant growth or ultimately to arsenic contamination of consumer products.

Literature Cited

1. Frost, D. V. Federation Proc. (1967), 26, 194.
2. Erlich, P., and Bertheim. Berichte (1907), 40, 3292.
3. Morehouse, N. F., and Mayfield, O. J. J. Parasitol., (1946), 32, 20.
4. Anonymous. "Feed Additive Compendium", The Miller Publishing Company, Minneapolis, Minnesota (1973).
5. Frost, D. V., Overby, L. R., and Spruth, H. C. J. Agr. and Food Chem. (1955), 3, 235.
6. Frost, D. V., and Spruth, H. C. Symposium on medicated feeds. Medical Encyclopedia, New York (1956).
7. Smith, L. W., Calvert, C. C., Frobish, L. T., Dinius, D. A., and Miller, R. W. USDA,ARS 44-224 (1971).
8. Peoples, S. A. Proc. Nat'l. Acad. Sci., Washington, D.C. (1969), 77.
9. Baron, R. R. Proc. Seminar at the Criterion, Lower Regent Street, London, S.W. 1, Salsbury Laboratories, London (1969).
10. Hansen, L. E., Hill, E. G., and Ferrin, E. F. J. Animal Sci. (1956), 15, 280.
11. Pope, C. W., and Schaible, P. J. Poultry Sci. (1959), 38, 350.
12. Russo, J. M., Hanson, L. E., and Jeseski, J. J. J. Animal Sci. (1954), 13, 998.
13. Vorhies, M. W., Sleight, S. D., and Whitehair, C. K. Cornell Vet. (1969), 59, 3.
14. Bridges, J. H., Hale, F., Kunkel, H. D., and Tyman, C. M. J. Animal Sci. (1954), 13, 912.
15. Moody, J. P., and Williams, R. T. Food Cosmet. Toxicol. (1964), 2, 687.
16. Moody, J. P., and Williams, R. T. Food Cosmet. Toxicol. (1964), 2, 695.
17. Moody, J. P., and Williams, R. T. Food Cosmet. Toxicol. (1964), 2, 707.
18. Frost, D. V. Poultry Sci. (1953), 32, 217.
19. Hansen, L. E., Carpenter, L. E., Aunan, W. J., and Ferrin, F. E. J. Animal Sci. (1955), 14, 513.
20. Overby, L. R., and Frederickson, R. L. J. Agr. and Food Chem. (1963), 11, 378.
21. Morrison, J. L. J. Agr. and Food Chem. (1969), 17, 1288.
22. Williams, K. R., and Whetstone, R. R. USDA Tech. Bull. 732, (1940).
23. Woolson, E. A. Paper No. 58, Pesticide Chemical Division Abstract. American Chem. Soc., Los Angeles, California (1974).

6

Microbiological Methylation of Arsenic

DONALD P. COX[1]

Department of Agronomy, Cornell University, Ithaca, N.Y. 14850

The biological methylation of arsenic was first recognized during several poisoning episodes in the early nineteenth century (1). Several individuals succumbed in their sleep to arsenic poisoning, the cause of which was not immediately recognized. Initially it was attributed to the presence of particles of arsenic in the room air originating from wallpaper pigments. Others thought biological activity might have reduced these arsenic pigments to arsine which was the toxic agent.

An experimental approach ultimately determined the cause of gas formation when an Italian scientist, Gosio, isolated and characterized some volatile arsenic-producing fungi. He exposed potato pulp containing arsenic trioxide to the air and soon detected growth of molds accompanied by a garlic-type odor. From these cultures he isolated a mold *Penicillium brevicaule* (*Scopulariopsis brevicaulis*) which produced copious quantities of gas with a garlic-like odor. Gosio trapped samples of the gas from this culture and converted it to arsenic proving that the metal had truly been volatilized by the fungus (2). An associate, Bignelli, performed an elemental analysis on the gaseous material and incorrectly deduced the structure of diethylarsine (3).

It wasn't for another thirty years that the structure of "Gosio gas" was determined by Challenger's group in England. A great deal of research on biological methylation of metals was performed by Challenger and associates over twenty odd years beginning with the methylation of arsenic compounds (1,4). These researchers cultured several strains of *S. brevicaulis* on bread crumbs containing arsenic trioxide and the gas produced was precipitated in aqueous solutions as either the mercuric chloride,

[1] Present address: Union Carbide Corp., Chemicals and Plastics, P. O. Box 8361, South Charleston, W. Va. 25303

picrate, or nitrate derivatives. These salts were tediously compared with derivatives made from authentic methylated arsenic compounds by melting point analysis. Challenger indisputedly proved the structure of "Gosio gas" to be that of trimethylarsine. From these studies, Challenger proposed a metabolic pathway in S. brevicaulis for the production of trimethylarsine from arsenite (Figure 1). To elucidate the mechanism of the methylation reaction, Challenger demonstrated that alkylated arsenicals supplied to the mold were methylated in available positions on the metal and were not first reduced to the inorganic state. This was shown by incubating ethylarsonic acid with the mold and isolating the volatile product, ethyldimethylarsine. This same experiment established that trimethylarsine could not be formed by dismutation or the movement of alkyl groups from one arsenical to another. A dismutation reaction would have resulted in a mixture of alkylated arsenicals (1,5).

Following Challenger's effort, little other experimentation concerning biological arsenic volatilization by pure cultures was performed for some time except for several notable observations. Thom and Raper (6) detected garlic odors produced by five species of Aspergillus, one of Penicillium and one species of Fusarium in the presence of 1500 µg/ml As_2O_3, however, the gas was not identified. Zussman et al. (7) observed volatilization of arsenate but not arsenite by Trichophyton rubrum.

The use of arsenicals as wood preservative agents spawned another study which revealed the biological methylation phenomenon (8). In this study a total of sixty-five species of "wood rotting" fungi were tested for inhibition by or metabolism of arsenic trioxide. Of these, two species of Lenzites provided evidence for volatile arsenic formation. The gaseous product presumed to be trimethylarsine was not identified. The study did not reveal whether arsenic-containing wood preservatives would promote a similar production of the arsenic gas by these two species.

Studies of arsenic methylation in soil have been also somewhat limited. The formation of volatile arsenic compounds by soil microorganisms was observed by Epps and Sturgis (9). When passing air through flooded soil containing arsenate, they were able to trap arsenic compounds in an acid solution receiving the exhaust air stream. Recent studies with an arsenic herbicide (cacodylic acid) further supported the hypothesis that arsenic-volatilizing organisms are ubiquitous in nature. Three types of soil demonstrated loss of total arsenic over several months of incubation. These samples containing high levels (100 µg/ml) of cacodylic acid emitted detectable garlic odors (10). Arsenic metabolism in the soil is discussed in greater detail in another portion of this symposium.

With the advent of ecological concern over disposal of waste arsenic compounds and the potential contribution of arsenic residues from pesticide application, new studies were initiated concerning biological methylation of arsenic. The biomass of waste treatment plants are a potential source of microorganisms which could form volatile arsenic compounds. Under anaerobic conditions, a species of *Methanobacterium* has already been observed to volatilize arsenic. Cell free extracts and/or whole cell preparations of this organism produced dimethylarsine in presence of arsenate or arsenite which were shown to obstruct the methane-forming mechanism (11). Under these conditions, methane formation was not only shown to be inhibited by arsenic compounds but also by selenium and tellurium compounds and previously by mercury (12).

Recognizing the potential for volatilization of arsenic via metabolism of pesticide residues, studies in our laboratory were begun to find organisms capable of methylating arsenic compounds. Aerobic enrichment cultures were initiated using domestic sewage and four different arsenic compounds. Sodium arsenite, sodium arsenate and salts of methylarsonic and dimethylarsinic acids were chosen on the basis of their use as pesticides and their being intermediates in Challenger's proposed pathway. After a short incubation period, garlic odors were noted in enrichment cultures containing either dimethylarsinic acid at pH 4, 5, and 7, monomethylarsonic acid at pH 5 or sodium arsenate at pH 4. Three different species of fungi were then isolated and identified as: 1) *Candida humicola* (Dazewska) Diddens and Lodder; 2) *Gliocladium roseum* Bain and 3) a *Penicillium* species (13). Each of these species was then examined for its ability to produce trimethylarsine from four different arsenic-containing compounds. The production of trimethylarsine by each culture appears in Tables I, II, and III. As is evident all three species formed the gas expeditiously on the methylated arsenic substrates at neutral and acid pH. Only the *Candida* species was able to demonstrate trimethylarsine production on inorganic arsenicals and only in acid conditions.

TABLE I (13)
Trimethylarsine Production by *Candida humicola*

Compound	nmole TMA in headspace		
	pH 5	pH 6	pH 7
DMA	87	41	2
MMA	9	6	0
$AsO_4^{\equiv}$	6	0	0
AsO_2^-	8	6	11

Culture age: 1 week, time after capping: 3 days.

TABLE II (13)
Trimethylarsine Production by *Gliocladium roseum*

Compound	nmole TMA in headspace		
	pH 5	pH 6	pH 7
DMA	10	52	253
MMA	2970	3700	2970
$AsO_4^{\equiv}$	0	0	0
AsO_2^-	0	0	0

Culture age: 1 week, time after capping: 6 days.

TABLE III (13)
Trimethylarsine Production by a *Penicillium* Species

Compound	nmole TMA in headspace		
	pH 5	pH 6	pH 7
DMA	22	14	6
MMA	429	62	254
$AsO_4^{\equiv}$	0	0	0
AsO_2^-	0	0	0

Culture age: 4 days, time after capping: 3 days.

The *Candida* species was unique in its ability to methylate four arsenicals and this organism was selected for more thorough studies of its control mechanisms (14). The greatest quantity of trimethylarsine per cell was produced when *C. humicola* was cultured in presence of either arsenate or dimethylarsinate (Figure 2). Smaller amounts were produced from cells grown in presence of arsenite or methylarsonate even though, in this case, their arsenic concentrations were about twice as high as in the arsenate or dimethylarsinate grown cultures. The specificity of the methylation reaction of this organism for arsenic compounds was next tested by growing cultures with other anions of Group 5 elements which were of similar chemical nature to arsenate. The phosphate ion is very similar in structure and chemical properties to the arsenate ion although stricter reducing conditions would be necessary to allow the formation of methylated phosphines. Not surprisingly, our attempts at viewing methylated phosphorus compounds formed by C. humicola metabolism were unsuccessful. Similarly, no gas production could be detected in presence of nitrate, nitrite, ammonia or antimonate ions.

The presence of some of these ions in a culture of C. humicola methylating arsenic was not, however, without effect. We did observe a definite inhibition by the presence of phosphate on trimethylarsine production from arsenate, monomethylarsonate or arsenite (Figure 3). The inhibition by phosphate of trimethylarsine formation from arsenate was effective at phosphate levels as low as 10 μg/ml or one one-hundredth the concentration of arsenate. Phosphate was completely inhibitory at 1000 μg/ml (0.1%) (Figure 4). The addition of phosphate to suspensions producing trimethylarsine from dimethylarsinate caused no inhibition (Figure 3) even at concentrations as high as eight times the arsenate concentration.

The observation that phosphate does not inhibit the conversion of dimethylarsinate to trimethylarsine by C. humicola is of special interest in light of the report by Da Costa (15) that phosphate did not relieve dimethylarsinate toxicity to arsenic-sensitive microorganisms. Phosphate addition did relieve arsenate and arsenite toxicity.

These two phenomena, arsenic toxicity and volatilization, might conceivably be coupled in mold metabolism. The toxicity mechanisms of arsenicals are not completely understood but one postulated mechanism involves their reaction with sulfhydryl groups in cellular protein. Another may be arsenate interference with oxidative phosphorylation, specifically inhibiting adenosine triphosphate synthesis. Possibly, a third mechanism of arsenic toxicity exists in which a single toxicant is formed from metabolism of the various arsenicals. Our results

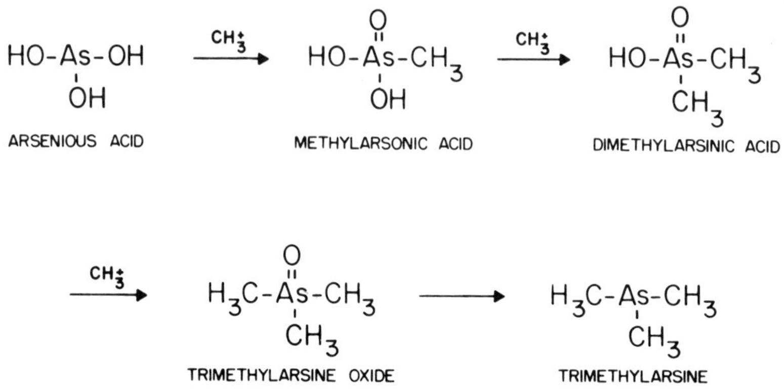

Figure 1. Challenger's proposed pathway

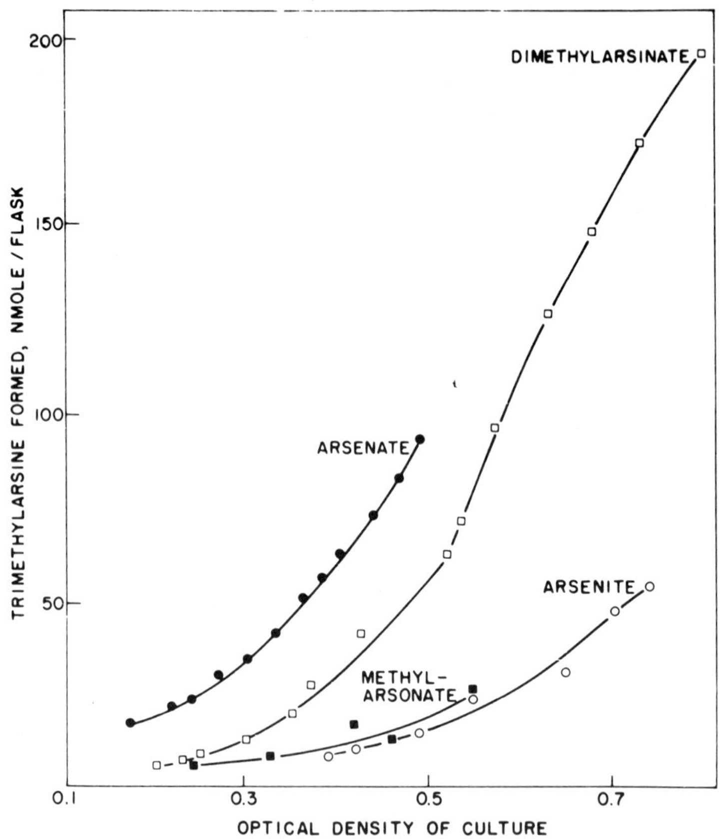

Figure 2. Formation of trimethylarsine from arsenate (240 μg As/ml), arsenite (544 μg As/ml), monomethylarsonate (577 μg As/ml), and dimethylarsinate (272 μg As/ml) in relation to cell density in growing cultures of Candida humicola

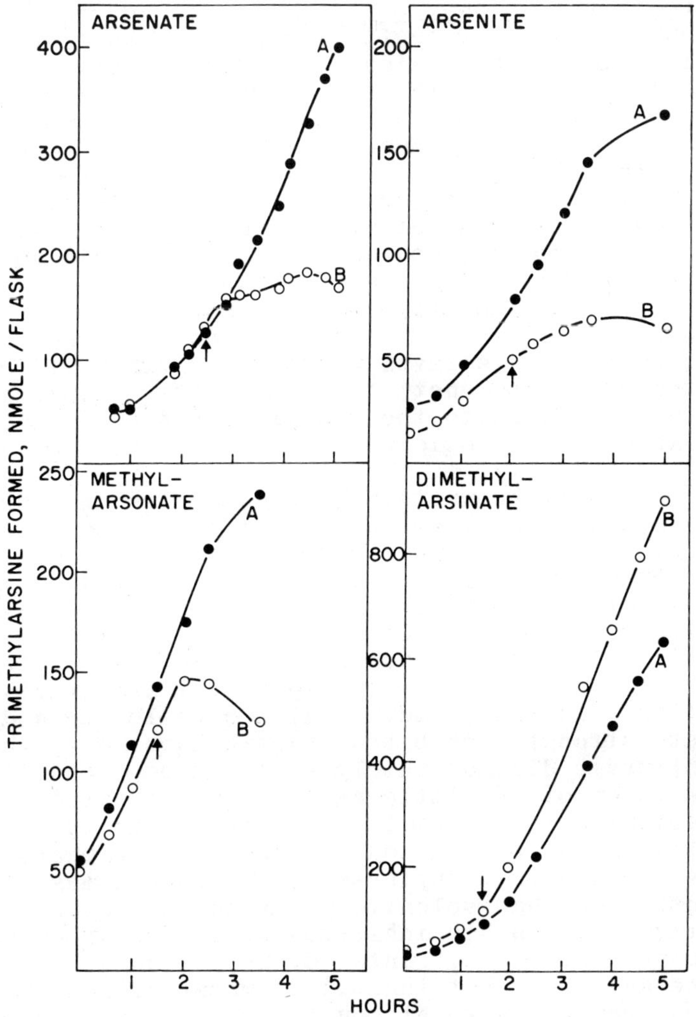

Figure 3. Effect of high phosphate levels on the formation of trimethylarsine by Candida humicola *grown in media with arsenate, arsenite, monomethylarsonate, and dimethylarsinate. The arrow indicates the time that supplemental* KH_2PO_4 *was added.*

A. No additions; B. 0.1% KH_2PO_4 *added*

indicated that phosphate suppresses gas evolution by preventing trimethylarsine formation in the pathway between monomethylarsonic and dimethylarsinic acids. The hypothetical inhibitor might then be an intermediate in the pathway of trimethylarsine production. The conversion of arsenate, arsenite or monomethylarsonic acid to the toxicant may conceivably be blocked by phosphate while dimethylarsinate conversion (if indeed, it is not the toxicant) to the hypothetical compound is not. Much further study is needed to clarify this point.

The inhibition of trimethylarsine formation is catalyzed with anions of other Group 5 elements. Surprisingly enough, sodium hypophosphite did not inhibit trimethylarsine as did orthophosphate (Figure 5). Antimonate was toxic to trimethylarsine production at equimolar concentrations to arsenate but nitrate at this level was not. Nitrite stopped trimethylarsine production but the site of inhibition could be unrelated to the methylation process since it is known to be toxic to cells at slightly higher concentrations.

It is well known that S. brevicaulis and certain other cultures can form methylated derivatives of selenium and tellurium compounds (1,11,16). Our strain of C. humicola was also shown to produce dimethylselenide from selenite and selenate and a volatile product from tellurate (17). It was determined if the presence of these anions and those of other Group 6 elements would affect trimethylarsine formation. In fact, sodium salts of selenite and selenate strongly inhibited trimethylarsine synthesis and tellurate did moderately so (Figures 6 and 7). The addition of as little as 10 µg/ml, a ratio of one part selenite to 100 parts of arsenate, inhibited trimethylarsine production (Figure 8). Selenate also inhibited trimethylarsine but a ten times higher concentration than selenite was required to be as effective. Tellurate inhibited at equimolar concentrations to arsenate. Interestingly enough, the sulfate anion did not inhibit trimethylarsine production from arsenate at equimolar concentrations of sulfate to arsenate (Figure 6). It was noted, however, that dimethylselenide production from selenite by C. humicola was partially inhibited by equimolar amounts of sulfate but not arsenate (Figure 9).

These previous results indicated that partially separate methylation systems in C. humicola might function for selenium and arsenic. To aid in studying this phenomenon, C. humicola was grown in presence of

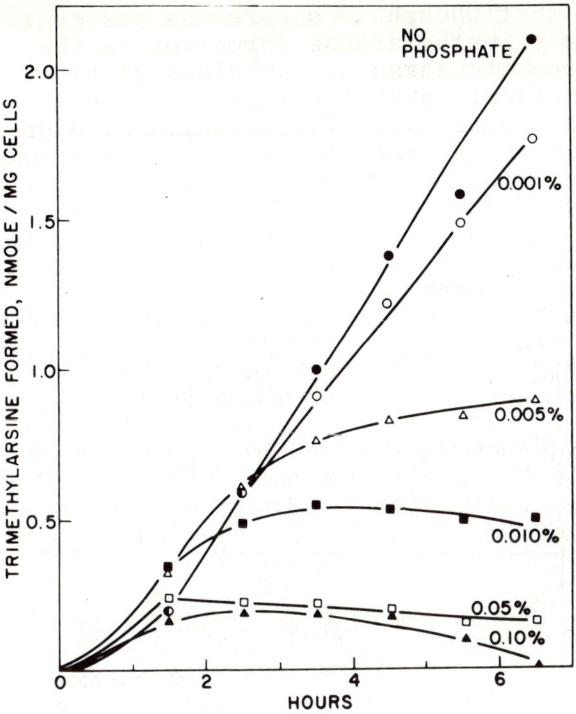

Figure 4. Inhibition of trimethylarsine formation from arsenate (240 μg As/ml) by resting cells in the presence of various KH_2PO_4 concentrations

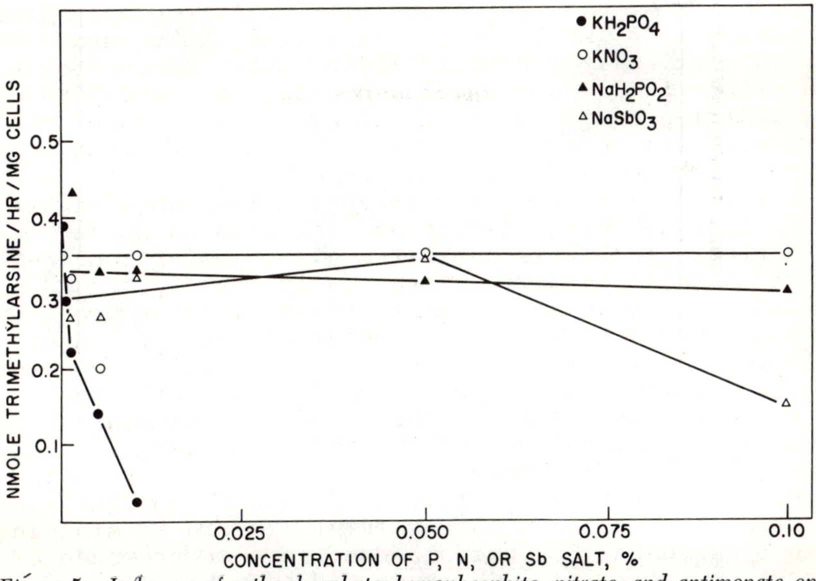

Figure 5. Influence of orthophosphate, hypophosphite, nitrate, and antimonate on the production of trimethylarsine by resting cells

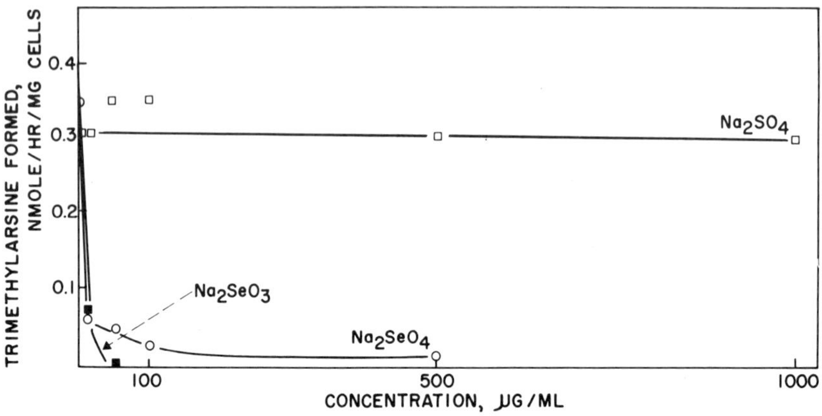

Figure 6. Trimethylarsine evolution from 1000 µg NaH_2AsO_4/ml by resting cells incubated with various concentrations of selenate, selenite, and sulfate

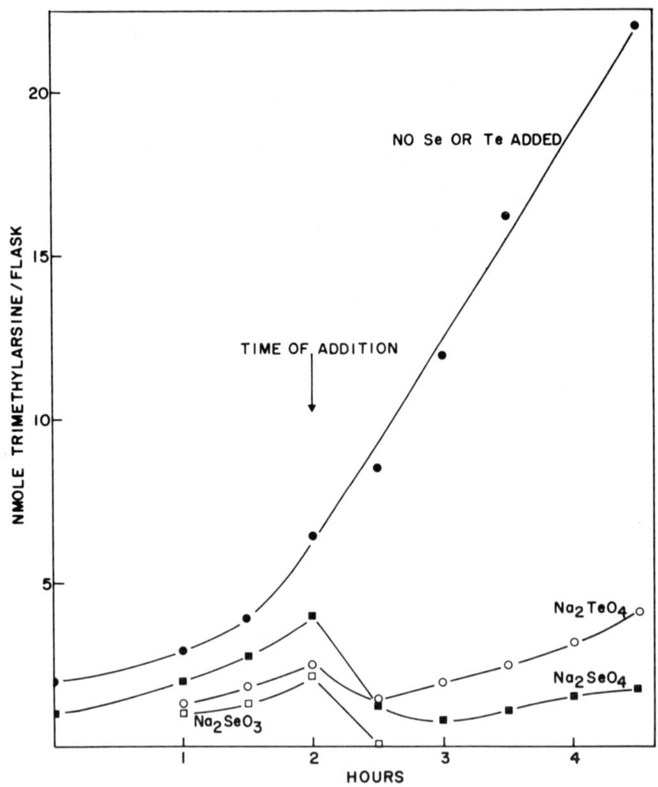

Figure 7. Trimethylarsine production by a growing culture of Candida humicola *in the presence of either 100 µg Na_2SeO_3, 1.0 mg Na_2SeO_4, or 1.0 mg Na_2TeO_4/ml*

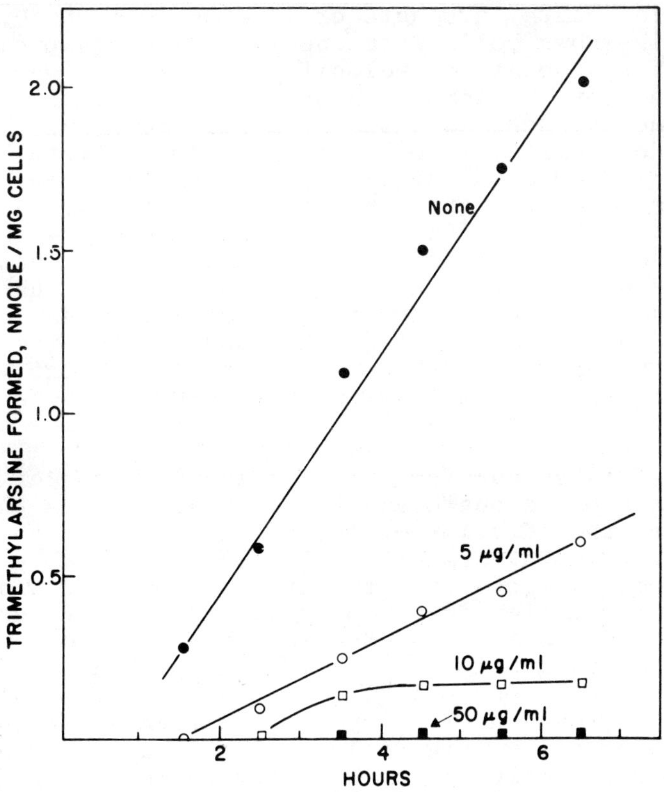

Figure 8. Selenite inhibition of trimethylarsine formation from NaH_2AsO_4 by resting cells of Candida humicola. *The concentrations shown are the levels of Na_2SeO_3.*

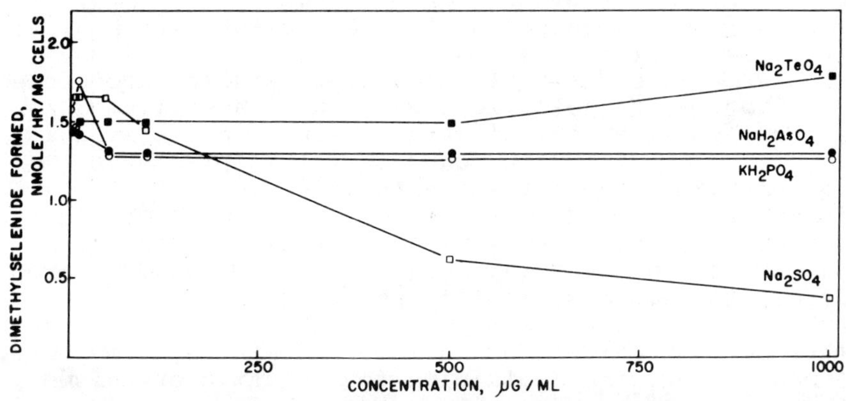

Figure 9. The effect of concentration of several compounds on the rate of dimethylselenide production by Candida humicola

either selenite, arsenate or in absence of both. The arsenate-grown cells were then incubated in presence of either arsenate or selenite and the time was measured for the formation of the appropriate gas. The same was done with the selenite-grown and the un-induced cells. The lag periods for trimethylarsine production from selenite-grown, arsenate-grown and un-induced cells were measured and compared. As is evident in Figure 10, the arsenate-induced cells produced trimethylarsine sooner than the selenite-induced cells which in turn were more rapid than the un-induced cells. Likewise, the selenite-induced cells produced dimethylselenide sooner than the arsenate-induced or the un-induced cells. One might conclude that the methylation systems have some separate enzymatic steps although more studies are needed.

As stated earlier, trimethylarsine production is inhibited by selenate and selenite as well as phosphate, however, the addition of these compounds did not only inhibit trimethylarsine production but actually appeared to remove trimethylarsine from a growing culture in a closed system as indicated in Figure 11. This was supported by demonstrating a loss of a known amount of authentic trimethylarsine placed in a closed system with selenite (17). A loss of the gas did not occur in presence of phosphate or arsenate or in absence of cells (Figure 12). Previous work describing a chemical complex which can form between selenium metal (18) and trimethylarsine may explain the phenomemon since *C. humicola* suspensions containing selenite produce an orange colloidal material, identified as selenium metal (17), in addition to dimethylselenide. No chemical evidence for such a trimethylarsine-selenium compound was found but if formed, could account for the severe inhibition of the arsenic-methylating system.

From an ecological standpoint, studies are needed to determine if volatile forms can be assimilated and stored by higher animals. An arsenic analysis on a species of marine kelp found that this plant was concentrating the metal one-thousand fold (19). In light of past evidence that methylated metals, namely mercury, can be bio-magnified to toxic levels by passing through natural food chains (20) studies are needed to determine if this phenomenon occurs with arsenic. To date one study has been made indicating that some arsenicals may not be as ominous as mercury. A model ecosystem containing four aquatic organisms with dimethylarsinic acid and dimethylarsine did not

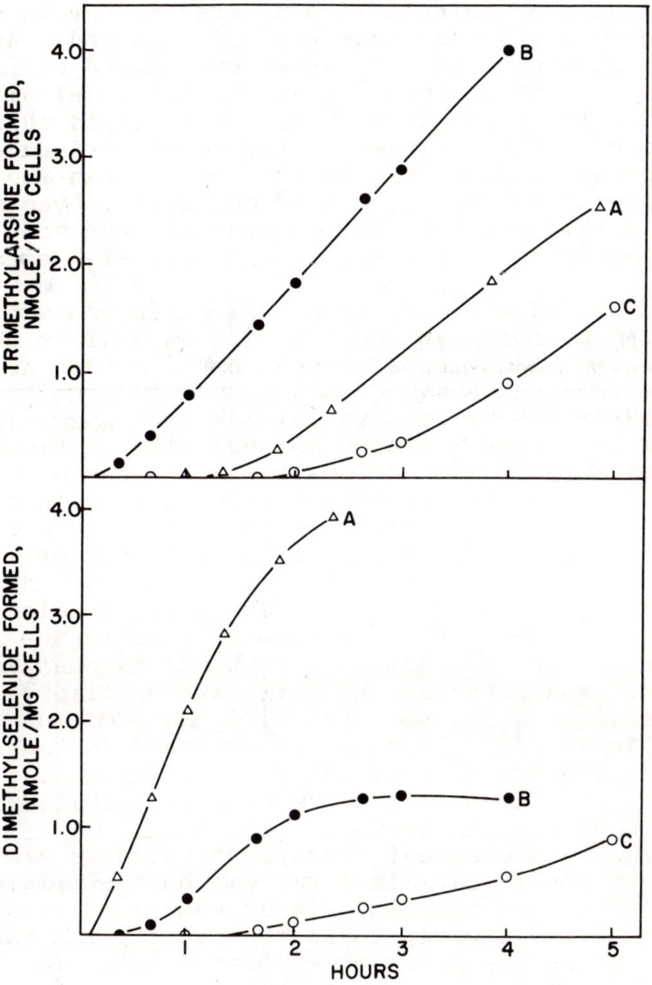

Figure 10. The formation of trimethylarsine (top) and dimethylselenide (bottom) by cell suspensions of Candida humicola *grown in the presence of 1.0 mg Na_2SeO_4/ml (A), 1.0 mg NaH_2AsO_4/ml (B), or neither (C)*

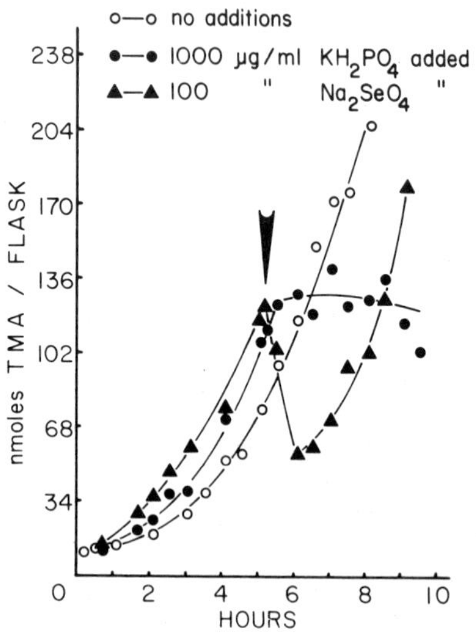

Figure 11. The effect of the addition of sodium selenate or potassium phosphate on trimethylarsine production by a growing culture of C. humicola

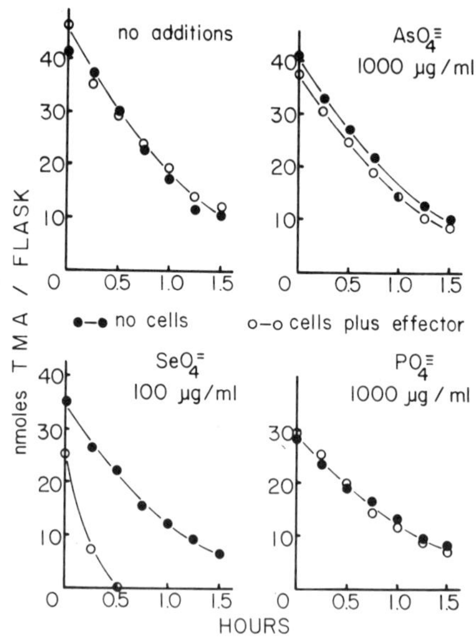

Figure 12. Rate of disappearance of trimethylarsine incubated with and without cell suspensions of Candida humicola *and 100 μg of Na_2SeO_4, 1.0 mg of KH_2PO_4, or NaH_2AsO_4/ml*

show significant bio-magnification (21).

In summary, evidence for the methylations of arsenic compounds by three mold species from sewage has been presented. The frequent use of arsenic compounds as pesticides provides a potential hazard if closed biological systems containing populations capable of methylating are exposed to arsenicals. One of these three species, a yeast, has been shown to methylate selenium and probably tellurium anions as well as arsenic compounds. The methylation mechanisms for selenium and arsenic in this organism appear to be partially found in a common pathway. Despite the widespread usage of arsenic compounds, little is known of the behavior of this element in natural ecosystems and with regard to the potential hazard of arsenic poisoning, the need for further study is self-evident.

Literature Cited

1. Challenger, F. Chem. Rev. (1945) 36:315-361.
2. Gosio, B. Arch. Ital. Biol. (1893) 18:253-265.
3. Gosio, B. Arch. Ital. Biol. (1901) 35:201-213.
4. Challenger, F. "Biological Methylation," Advances in Enzymology, pp. 429-491, Interscience Publishers, New York, 1951.
5. Challenger, F., Higginbottom, C., and Ellis, L. J. Chem. Soc. (1933) 1933:95-101.
6. Thom, C. and Raper, K. Science (1932) 76:548-550.
7. Zussman, R., Vicher, E. and Lyon, I. J. Bacteriol. (1961) 81:157.
8. Merrill, W. and French, D. W. Proc. Minn. Acad. Sci. (1964) 31:105-106.
9. Epps, E. A. and Sturgis, M. B. Proc. Soil Sci. Soc. Amer. (1939) 4:215-218.
10. Woolson, E. A. and Kearney, P. C. Environ. Sci. Tech. (1973) 7:47-50.
11. McBride, B. C. and Wolfe, R. S. Biochemistry (1971) 10:4312-4317.
12. Wood, J. M., Kennedy, F. S., and Rosen, C. G. Nature (London) (1968) 220:173-174.
13. Cox, D. P. and Alexander, M. Bull. Environ. Contam. Toxicol. (1973) 9:84-88.
14. Cox, D. P. and Alexander, M. Appl. Microbiol. (1973) 25:408-413.
15. Da Costa, E. W. B. Appl. Microbiol. (1972) 23:46-53.
16. Fleming, R. W. and Alexander, M. Appl. Microbiol. (1973) 24:424-429.

17. Cox, D. P. and Alexander, M. Microbial Ecology (1974) 1 (in press).
18. Renshaw, R. R. and Holm, G. E. J. Amer. Chem. Soc. (1920) 42:1468-1471.
19. Gorgy, S., Rakestraw, N. W., and Fox, D. L. J. Marine Res. (1948) 7:22-32.
20. Katz, A. Crit. Rev. Environ. Centr. (1972) 2:517-534.
21. Isensee, A. R., Kearney, P. C., Woolson, E. A., Jones, G. E. and Williams, V. P. Environ. Sci. Tech. (1973) 7:841-845.

7

Bioaccumulation of Arsenicals

E. A. WOOLSON

Pesticide Degradation Laboratory, Agricultural Research Service,
U.S. Department of Agriculture, Beltsville, Md. 20705

Arsenic, as a natural component of aquatic systems, is accumulated by aquatic organisms. The bioaccumulation ratio (BR) for arsenic, as used in this paper, is defined as the ratio of arsenic concentration in the organism divided by the concentration of arsenic in water. An arsenic content of 2 μg/liter or 2 ppb, an approximate average for sea water, will be assumed for data supplied by authors who did not measure the concentration of arsenic in sea water when analyzing arsenic in marine organisms.

Salt water (marine) organisms accumulate much more arsenic than do most fresh water organisms. Furthermore, aquatic plants accumulate much more arsenic than do higher members in the aquatic food chains. In marine plants, BR values of 71,000 have been observed.

A. Arsenic in fresh waters

Arsenic concentrations measured in fresh water bodies are listed in Table 1. The concentrations reported in a survey of U. S. lakes and rivers ranged from <10-1100 μg/liter (1). The high value of 1100 μg/liter was taken downstream from a sewage treatment plant on the Irwin Creek in North Carolina. The contamination source was never identified although a company producing arsanilic acid had been in operation recently. At the next monitoring station downstream, the arsenic level returned to 10 μg/liter. Arsenic is precipitated and complexed in water and appears in the sediments of both lakes and rivers. Searles Lake, California, a highly saline lake, had the highest values of any water reported. Other high saline lakes, however, contain only traces of arsenic. The Waikato River in New Zealand has arsenic present from geothermal waters, the outflow from Lake Taupo, the Wairakei thermal area, and the thermal region at Orakei-Korako. Spring waters (2) which were high in bicarbonate and passed through thermal layers frequently contain elevated arsenic levels.

Table 1. Arsenic content of natural fresh waters

Source	Concentration (μg As/l)	Reference
U. S. Lakes, Wisconsin	4-117	(3)
" " , Michigan	0.5-2.4	(4)
" " , Superior	0.1-1.6	(4)
" " , Searles Lake Calif.	198,000 - 243,000	(5)
" " , Calif.; <2000 ppm dissolved solids	0-100	(2)
" " , Calif.; >2000 ppm dissolved solids	0-2000	(2)
" lakes and rivers	<10-1100	(1)
" Columbia River	1.6	(6)
Japan, lakes	0.16-1.9	(6)
" , rivers	0.25-7.7	(6)
Germany, Elbe River	20-25	(6)
Greece, lakes	1.1-54.5	(6)
Sweden, rivers	0.2-0.4	(6)
N. Zealand, Waikato River[a/]	5-100	(8)
Spring Waters[b/]; California, Kamchatka, USSR, N. Zealand	130-1000	(2)
Oil and gas field waters; California, Louisiana, Hungary	0.0-5800	(2)
Thermal waters; Wyoming, Nevada, California, Alaska, Iceland	20-3800	(2)
Spring waters[c/]; USSR, Wyoming, Algeria, Iceland	30-500	(2)

a/ High in bicarbonate, geothermal origins
b/ High in bicarbonate and boron
c/ Deposit travertine

High arsenic levels have been found in some drinking waters (Table 2). Contamination arises from obtaining water that passes through layers of arsenical pyrites, leaching of mine waste piles (9), or drilling wells through layers of contaminated soils (10). The last-named source of contamination resulted from improper storage of an arsenical grasshopper bait, and those drinking the water became ill. Of the 18,204 community water supplies studied, fewer than 1% exceeded the recommended 10 μg/liter level for arsenic (14). The average content of arsenic in fresh water is below the recommended drinking water quality standards for the U. S. and probably is about 1.5 to 2 μg/liter.

Table 2. Arsenic content of drinking waters

Source	Concentration (μg As/1)	Reference
Argentina	tr-300	(11)
Argentina, Cordoba Province	300-1000[a]	(12)
Canada	300-7500	(13)
U. S.	tr-100[b]	(14)
" , Wenatchee, Wash.	5-6[c]	(15)
" , Minnesota	11,800-21,000[d]	(10)
Taiwan	250-850[e]	(16)

a/ Contains vanadium also; contains waste products from mining
b/ Less than 1% exceeded 10 μg/1
c/ Watershed which received lead arsenate
d/ Well dug through area containing grasshopper bait
e/ Naturally contaminated artesian well water

B. Arsenic in marine waters

Concentrations of arsenic in the marine environment generally range from 0.15 to 6 μg/liter, with an average value of ca. 2 μg/1 (Table 3). Areas near the river outflows tend to have higher arsenic levels than the oceans as a whole. The data of Chapman (17) are 100-fold higher than those reported by other authors. No apparent reason for this discrepancy exists.

Scientists disagree as to the oxidation state of arsenic in sea water. Many believe that arsenic exists in sea water in the +3 state; others believe that it is predominately in the +5 state. Sugawara and Kanamori (18) showed that the +5 As/total As ratio was close to 0.8 in ocean water. Diagrams of E_h vs pH for sea water indicate that arsenic would be in the +5 state in all oxidizing layers of the ocean (19). Chemical reduction may take place at lower depths where oxidation is not great or may be mediated by microorganisms. In any event, an equilibrium between arsenate and arsenite exists in sea water. The same type of equilibrium probably exists in any stratified lake (19) where precipitation, oxidation and reduction, and adsorption, reduction and methylation occur within a single lake.

In the aerobic aqueous layer, reduced forms of arsenic will be oxidized to arsenate and both may coprecipitate with ferric hydroxide. Turbulence and convection transport arsenate to the oxygen-depleted hypolimnion where it may be reduced and form arsenite or AsS_2. These reactions depend on sulfur concentration and the E_h. Coprecipitation, absorption, adsorption, and crystal

Table 3. Arsenic content of marine waters

Source	Concentration (μg/As/l)	Reference
England	106-760	(17)
England	2	(20)
English Channel	2-5	(21, 22)
Pacific coast	3-6	(6)
N. W. Pacific Ocean	0.15-2.5	(6)
Indian Ocean	1.3-2.2	(6)
S. W. Indian Ocean	1.4-5.0	(6)

growth may cause arsenate to accumulate in the sediments, where reduction of ferric arsenate and arsenite results in either solubilization as arsines, stabilization as insoluble sulfides, or reduction to arsenic metal. Microbial methylation and reduction solubilize arsenic, also. Further, diffusion through the sediments, or mixing by currents or burrowing organisms, can cause the arsenic to reenter the water phase.

Arsenic in the water phase is adsorbed or incorporated into benthic organisms, algae, zooplankton and phytoplankton. This arsenic is converted to organo-arsenical compounds as well as water-soluble compounds within these organisms (23, 24, 25). Fish consume the algae or microscopic organisms and further transform the arsenical to a more complex compound (26). Crustacea and filter-feeding shellfish may absorb arsenic from the water directly or from microscopic organisms.

C. Arsenic in fresh water organisms

In model ecosystem studies (Table 4), catfish and *Gambusia* concentrated arsenic from <1 to 130 times the conc. in water respectively. Daphnids and crayfish concentrated arsenic by factors of 4 and 5, respectively, in water, while algae concentrated it to 34 times the water concentration. Concentrations of cacodylic acid and arsenate in fresh water organisms do not appear to correlate well with one another. Algae and daphnids accumulated more cacodylic acid than they did arsenate, but *Gambusia* fish accumulated much less. MSMA (monosodium methanearsonate) accumulated to give BR values similar to those for arsenate, except the value was ninefold less for algae.

The ^{14}C-BR values for ^{14}C-cacodylic acid (CA) in the aquatic system were much higher than those for arsenic, possibly because it degraded. ^{14}C-BR values reached 163 to 27,000 in algae, 89 to 35,000 in duckweed, 4 to 1,000 in snails, 2 to 275 in catfish, and 3 to 14 in crayfish (27). Corresponding BR values for

Table 4. Bioaccumulation Ratio (BR)[a/] values for As obtained in model ecosystem studies

Compound tested	Water conc. ppm As	Species							Reference
		algae	duckweed	*Daphnia*	snail	*Gambusia*	catfish	crayfish	
$^{74}AsO_4$	1	34	ND[b/]	4	ND	130	ND	5	(*26*)
^{14}C-MSMA	1	4	ND	8	ND	108	ND	1	(*26*)
^{14}C-CA	~1	<3-17	1-3	ND	<1-23	--[c/]	<1	<1-16	(*27*)
	0.1	45	-	39	9-20	<1	-	-	(*28*)
	1.0	17	-	42	2-68	<1	-	-	(*28*)
	10.0	7	-	25	1-7	<1	-	-	(*28*)

a/ BR = conc. in tissue/conc. in water

b/ Not Determined

c/ Organism not in the experiment

arsenic in this same experiment were: algae, 3 to 17; duckweed, 0 to 3; snails, 2 to 21; catfish, <1; crayfish <1 to 16 (28). Presumably, the ^{14}C is split from the As atom and incorporated as labeled $^{14}CO_2$ or some other metabolite, possibly CH_3.

D. Arsenic in aquatic animals

Concentrations of arsenic are generally much lower in fresh water fish than in marine fish (Table 5). A BR value of 3 to 30 for various fresh water fish was reported by Pratt et al. (29) and 10 to 20 by Gilderhus (30). Marine organisms concentrated arsenic to give BR values ranging from 9 for crustacea and shellfish (17) to 64,100 for shrimp (31). The latter value was on a dry weight basis. Dry weight contents and the resulting ratios are always higher than those calculated on a fresh weight basis.

Table 5. Bioaccumulation Ratio values for As in aquatic organisms[a]

Specie	ppm As in tissue	BR values[b]	Reference
Haddock	2-10.8	1000-5400[c]	(5) (25)
Kingfish	8.86	4430	(5)
Crustacea and	1.5-3.1	750-1550	(5)
shellfish	0.018-1.06	9-530	(17)
Assorted fish	0.076-2.27	38-1135	(32)
Assorted fish	<1-6.4	<500-3,200[c]	(33)
Shrimp	3.6-48	1836-64,100[c]	(31) (25)
Mackrel	4.7-9.2	2350-4600[c]	(25)
Cod	24.3	12,150[c]	(25)
Assorted fresh	0.1-0.2	10-20	(30)
water fish	0.035-0.298	3-30[d]	(29)

[a] Concentration in tissue/concentration in water
[b] A marine concentration of 2 μg As/l is used in all calculations
[c] Dry weight basis
[d] Water concentration assumed to be 10 μg As/liter

In general, for various assorted salt water fish, the concentration ratios are 10 to a 100 times higher than those reported for fish and crustacea in fresh water. This difference may be a function of arsenic concentration to which the organism is exposed on a continuing basis. Arsenic concentrations are generally lower in fresh water than in the marine environment.

E. Arsenic in aquatic plants

Bioconcentration ratio values for arsenic in aquatic fresh

water plants ranged from 1 for duckweed in a model ecosystem study (27) to 20,000 (7) for submerged weeds in the Waikato River in New Zealand (Table 6). The arsenic in the Waikato River is from geothermal origins and at places is quite high. Emerged weeds from the same river concentrated arsenic only about 100 times. Lakeweeds concentrated arsenic to give BR values of 110-14,500 (8). Cacodylic acid was accumulated by algae to a much higher degree than arsenate (28), whereas MSMA BR values were about the same (26) as for arsenate.

Table 6. BR values for arsenic in fresh water plants[a]

Specie	ppm As in tissue	BR values	Reference
Algae	2-12	3-17	(27)
Algae	27	34	(26)
Algae	4	4[b]	(26)
Algae	4.5-9.8	7-1635[c]	(28)
Algae	-	590-4600	(24)
Algae	550	7,000	(7)
Submerged weeds	20-971	800-20,000	(7)
Emerged weeds	8-12	100	(7)
Lakeweeds	11-1450	110-14,500	(8)
Duckweed	1-3	1-3	(27)

a/ BR = concentration in plant/concentration in water
b/ Treated with MSMA
c/ Treated with cacodylic acid

Marine plants accumulated or bioconcentrated arsenic to a much higher level than fresh water plants (Table 7). Seaweed contained 350 to 71,000 times more arsenic than sea water (34, 25). Algae contained between 50 and 47,500 times the concentration in water, depending on the species.

Table 7. BR values for arsenic in marine plants[a]

Specie	ppm As in tissue	BR values	Reference
Algae	4.8-94	2,400-47,000	(36)
Algae	0.1-95	50-47,500	(37)
Algae	1.67	835	(20)
Algae	--	200-3000	(24, 35)
Seaweed	0.7-12	350-6000	(5)
Seaweed	60-142	30,000-71,000	(25, 34)

a/ Concentration in tissue/concentration in water

Since arsenicals do bioaccumulate in the aquatic environment, the question of possible biomagnification exists. Biomagnification is the process whereby succeeding organisms in the food chain have higher levels of a contaminant than previous members. With arsenic, this does not appear to be true. Algae, which are low in a food chain, contain much higher arsenic levels than do fish and consumers of fish. Since the BR value is derived by dividing the arsenic concentration in tissue by the arsenic content in water, one reason for high BR values in algae is the inherently low value for arsenic both in sea water and in fresh water. Algae does concentrate arsenic against a concentration gradient. The actual levels of arsenic in any organism seldom exceed 100 ppm.

F. Form and toxicity of incorporated arsenic

Two further questions about arsenic bioaccumulation are the identity of the arsenic compound after incorporation and its resulting toxicity. Lunde (23, 24, 25, 35) and Woolson et al. (26) have shown that arsenic is present in algae and fish as both water-soluble and lipid-soluble arseno-organic compounds. Lunde concluded that "algae seem to be an important source of the arseno-organic compounds found in higher marine organisms." In studies with fresh water model ecosystems (26), small amounts of arsenate were present in all organisms fractionated on a Sephadex column (Fig. 1). However, algae contained an additional unknown, X_1, and crayfish contained another unknown X_2, in addition to X_1. MSMA (Fig. 2) in the same ecosystem was found in all three organisms and was apparently transformed to another unknown, X_3, only in *Daphnia*. About 30% of the MSMA was transformed, based on ^{14}C contents of the extracts. These studies by Lunde and Woolson indicate that arsenate is not a major form of arsenic in aquatic organisms.

The toxicity to sheep of arsenic in lakeweeds has been examined by Lancaster et al. (8). The weed, *Lagarosiphon major*, contained 288 ppm As, and was fed as 20% of the sheep diet for 3 weeks with no ill effects on the animals' health. Residues increased during feeding, but declined when the lakeweed was removed from the diet.

Coulson (31) fed shrimp, containing 128 ppm As, to rats at a dietary level of 13.3 ppm As. He also included As_2O_3 in his feeding studies at the same dietary level. The rat livers contained 20-fold less arsenic with the shrimp diet than with As_2O_3 in the diet. More than 98% of the arsenic fed in the shrimp was excreted within 4 days of feeding, whereas only 21% of the As_2O_3 fed was excreted. When humans ate shrimp, all of the arsenic was excreted within 4 days after consumption. The arsenic contained in shrimp obviously is not as toxic or as easily absorbed as As_2O_3.

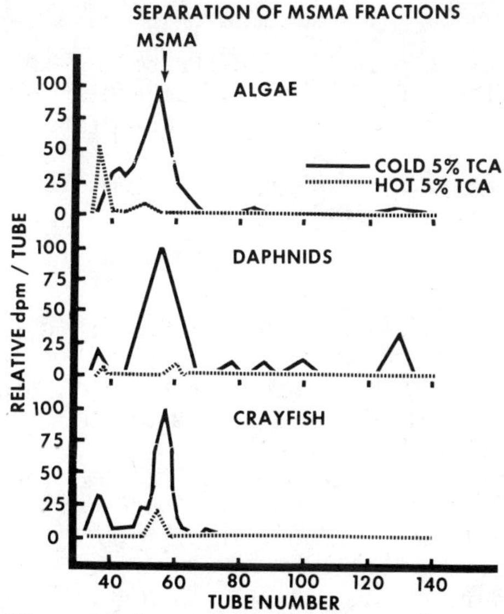

Figure 1. Separation of incorporated arsenic compounds from aquatic organisms treated with $^{74}As_4$ in a model ecosystem. X_1—tubes 75–80; X_2—tubes 90–110

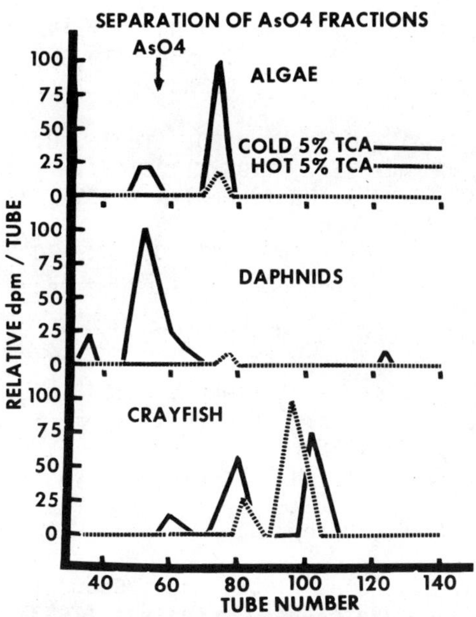

Figure 2. Separation of incorporated arsenic compounds from aquatic organisms treated with ^{14}C-MSMA in a model ecosystem. X_3—tubes 120–135

G. Conclusions

In summary, arsenic is bioconcentrated by aquatic organisms but not biomagnified. Plants usually accumulate more arsenic than fish, and crustacea accumulate intermediate amounts. Marine organisms normally contain more arsenic than their fresh water counterparts. However, the arsenic contained in the organisms is apparently not toxic to animals or humans, and is readily excreted.

Literature Cited

1. Durum, W. H., Hem, J. D., Heidel, S. G., Geo. Survey Circ., (1971), 643.
2. Fleisher, M., "Data of Geochemistry", 6th edit. Geo. Survey Prof. Paper 440, (1963), U. S. Govt. Printing Office, Washington, D. C.
3. Chamberlain, W., Shapiro, J., Limnol. Oceanogr., (1969), 14:921-7.
4. Seydel, Ilse S., Archiv fur Hydrobiologie (1973), 71(1): 17-30.
5. Schroeder, H. A., Balassa, J. J., J. of Chronic Diseases (1966), 19:85-106.
6. Wedepohl, K. H., "Handbook of Geochemistry", (1969), Vol. II, Ch. 33, Springer-Verlag, Berlin.
7. Reay, P. F., J. Appl. Ecol., (1972), 9(2):557-65.
8. Lancaster, R. J., Coup, M. R., Hughes, J. W., N. Z. Vet. J. (1971), 19:141-45.
9. Neubauer, O., Br. J. Cancer (1947), 1:192-251.
10. Feinglass, E. J., N. Eng. J. Med. (1973), 288:828-30.
11. Bado, A. A., J. Am. Water Works Assn., (1937), 31:1975-1977.
12. Arguello, R. A., Cenget, D. D., Tello, E. E., Rev. Argent, Dermatosif., (1938), 22:461.
13. Wyllie, J., Can. J. Publ. Health (1937), 28:128.
14. McCabe, L. J., Symons, J. M., Lee, R. D., Robeck, G. G., J. Am. Water Works Assn. (1970), 62:670-87.
15. Fairhall, L. T., New Eng. Water Works Assn. (1941), 55: 400-10.
16. Tseng, W. P., Chu, H. M., How, S. W., J. Nat'l. Cancer Inst. (1968), 40:453-63.
17. Chapman, A. C., Analyst (1926), 51:548-63.
18. Sugawara, K., Kanamori, S., Bull. Chem. Soc. Japan (1964), 37(9):1358-63.
19. Ferguson, J., Gavis, J., Water Research (1972), 6:1259-74.
20. Portmann, J. E., Riley, J. P., Anal. Chim. Acta (1964), 31:509-19.
21. Smales, A. A., Pate, B. D., Anal. Chem. (1952), 24:717.
22. Smales, A. A., Pate, B. D., Analyst (1952), 77:188.
23. Lunde, G., Nature (London) (1969), 224(5215):186-7.

24. Lunde, G., Acta Chemica. Scandinavica (1973), 27:1586-94.
25. Lunde, G., J. Sci. Fd. Agric. (1973), 24:1021-27.
26. Woolson, E. A., Kearney, P. C., Isensee, A. R., McShane, W. G., Irgolic, K. J., Zingaro, R. A., 168th ACS Meeting (1974), Paper No. 8, Pest. Chem. Div.
27. Schuth, C. K., Isensee, A. R., Woolson, E. A., Kearney, P. C., J. Ag. Food Chem. (1974) in press.
28. Isensee, A. R., Kearney, P. C., Woolson, E. A., Jones, G. E., Williams, V. P., Environ. Sci. & Tech. (1973), 7(9):841-5.
29. Pratt, D. R., Bradshaw, J. S., West, B., Utah Academy Proceedings (1972), 49(1):23-26.
30. Gilderhus, P. A., Trans. Am. Fisheries Soc. (1966), 95(3):289-96.
31. Coulson, E. J., Remington, R. E., Lynch, K. M., J. Nutrition (1935), 10:255-70.
32. Cox, H. E., Analyst (1925), 50:3-13.
33. Windom, H., Stickney, R., Smith R., White, D., Taylor, F., J. Fish Res. Board Can. (1973), 30:275-9.
34. Wilson, S. H., Fieldes, M., N. Z. J. of Sci. & Tech. (1941), 23B:47-48.
35. Lunde, G., Acta Chemica Scandinavica (1972), 26:2642-44.
36. Jones, A. J., Chemist and Druggist (1922), 97:166-7.
37. Vinogradov, A. P., Sears Found. Mar. Res. (1953), Memoir 2.

8

Arsenic in the Environment

ROBERT S. BRAMAN

University of South Florida, Tampa, Fla. 33620

Introduction

Environmental studies of arsenic are intriguing exercises largely because (1) the chemistry of this element provides for several molecular forms to exist in both air and water; (2) interactions with the biosphere can occur; and (3) in most instances analytical chemistry is required involving very small sample sizes. A brief historical excursion into the subject of arsenic in the environment would start with the ancient, known toxicity of compounds of this element, touch on the toxicity of arsenic if used in wallpaper and be composed thereafter of laboratory studies on the biomethylation of arsenic in cultures, analyses for total arsenic, analyses of industrial locations for arsine in air, and radiochemical tracer studies of arsenic in soil experiments. Much of this work has gone a long way in helping to define the analytical chemistry problem, a logical starting point in studying the environmental chemistry of arsenic.

Because of the likelihood that methylated forms and inorganic forms of arsenic would exist, it became obvious early that analytical methods were needed to speciate arsenic, that is, to determine the chemical forms present and at concentrations ordinarily found in the environment. Except in instances of high arsenic pollution, total arsenic in water is in the 1-10 ppb range, and in air in the 0-10 ng/m^3 range. Prior to work in our laboratories adequate methods for air and water arsenic speciation analyses at these levels had not been developed. The problem of analyzing water, air and to some extent biological samples for methylarsenic compounds and inorganic As(III) and As(V) at environmental concentrations usually encountered has now to a considerable extent been solved. By using these methods and the laboratory work of others it has been possible to better substantiate the environmental chemistry of arsenic.

Analytical Methods

Many methods have been developed for the determination of total arsenic and for the several arsenic compounds of interest in environmental studies. The silver diethyldithiocarbamate method (1) has a limit of detection of 0.2 μg As or 2 ppb, too insensitive for much work. In addition, the methylarsenic compounds produce a much different molar absorptivity than does inorganic arsenic (2). Thus the method is inaccurate. Peoples, Lakso and Lais (3) made use of this in developing a differential spectrophotometric method for partially distinguishing between methylarsenic and inorganic arsenic compounds down to 0.02 ppM.

Neutron-activation and atomic absorption methods have been widely used for total arsenic. Talmi and Feldman (4) are developing gas chromatographic-emission detection methods and have reported their progress in this symposium.

Johnson and Pilson (5) have developed a colorimetric method for distinguishing between arsenic (III) and arsenic (V) and applied it to sea water analyses. The method is not suitable for organic analyses.

Based upon work in our laboratories (2, 6, 7) methods have been developed and applied to water analyses for the speciation of arsenic in aqueous samples. Fundamentally, the method involves reduction of methylarsenic acids or inorganic arsenic to the corresponding arsines which are then separated and detected in an emission type detection system. Observation of the arsenic emission line gives a limit of detection well below 1 ng and a positive means of identifying arsenic. Identification of methylarsenic compounds in samples has been confirmed by comparison to standards and by a modification of the colorimetric method developed by Peoples, Lakso and Lais (3). Identification is also supported by the reduction pH selectivity of the several arsenic compounds. The method has been extensively studied (2) and also successfully used by others (8) in environmental analyses. It has been adapted for use to analysis of biological materials (9).

Air analyses for arsenic have proved difficult. Arsenic is present as fine particulate (8). Gordon and Zoller (10) have found that the element is enriched in particulate far above that expected for average coastal materials. This again implied inefficient collection of very fine particulate or the presence of vapor forms.

Because of known environmental methylation of arsenic from our earlier work on water analysis it was decided to attack air analysis from the point of view that trimethylarsine, dimethylarsine, methylarsine, arsine and their oxidation products as well as inorganic, non-volatile As(III) and As(V) could be present in air and had to be specifically identified. The analytical method for water analysis provided a good procedure to apply so long as sample collection and processing could be suitably adapted. Early work with solution air scrubbers and Millipore filters showed that the volatile forms of arsenic are not trapped quantitatively (if at all) and that, moreover, glassware was an intolerable contributor to inorganic arsenic background. Studies were then initiated on collection of the volatile arsines on gold-coated and silver-coated glass tubes of the same type used in earlier, successful air analyses for volatile forms of mercury (11).

Arsines were quantitatively but irreversibly absorbed onto gold-coated glass beads. Silver-coated glass beads did, nevertheless, prove to be a satisfactory collecting medium. All of the volatile arsines were found to be quantitatively collected on 4 cm long columns at air flow rates up to at least a face velocity of 32 cm/second from 22°C to 35°C and under conditions of up to 100% relative humidity.

Quantitative removal of the arsines without dealkylation or disproportionation was obtained by a mild, warm alkaline wash followed by a hot water wash. The arsines are probably air oxidized in the washing process since analysis of the air sample collecting tubes required reduction by sodium borohydride to detect the methylarsines. A typical analysis is shown in Figure 1. Identification of peaks as trimethylarsine or dimethylarsine again was done by retention time comparison with known standards. The limit of detection for air analysis was approximately 0.3-0.5 ng per sample. Concentration limits of detection obviously depend upon sample size.

These methods briefly described above have been used to carry out environmental studies on arsenic. Results of this together with the referenced works of others has helped to definitely establish the ambient environmental chemical forms of arsenic and to better define its chemical transformations.

Environmental Forms of Arsenic

Table I summarizes the molecular forms reported as found in the environment. Table II lists the analytical results on a number of samples. Not included are commercial phenylarsenic type compounds, obviously present where used, but not yet identified remote from such locations. Methyl and dimethylarsinic acid or salts are not excluded from the table because they are produced by biomethylation and are not exclusively commercial products. It is suspected that trimethylarsine and possibly dimethylarsine may exist to an unknown extent as a complexed or oxide form in air or in water but this has not been definitely established.

The listing of the methylarsenic compounds does not imply that they are completely ubiquitous. For example, many water samples have been analyzed giving less than the limit of detection of methylarsenic compounds. Sea water is a case in point. In work done at several locations in the Sargasso Sea (9), methylarsenic acids were not detected in open ocean water. This means that if present they were less than 1% of the total sea water arsenic which is approximately 2-3 ppb.

TABLE I. ENVIRONMENTAL FORMS OF ARSENIC

Compound	Source	Reference
	WATER	
As(III), arsenite ion and As(V), arsenate ion	Sea water	5, 6, 8, 9
	Fresh water ponds, rivers, lakes	6, 8
$CH_3AsO(OH)_2$	Sea water	6, 9
	Fresh water ponds, rivers, lakes	5
$(CH_3)_2AsO(OH)$	Sea water	6, 9
	Fresh water	6, 8
$(CH_3)_3As$ (or the oxide)	Fresh water	*
	AIR	
As(III) and As(V)	Particulate	8
CH_3AsH_2	Over As-treated soil	*
$(CH_3)_2AsH$	Over-treated soil	* 14, 15, 17
$(CH_3)_3As$	Over-treated soil	* 14, 15, 17

*Reported in this article

TABLE I. ENVIRONMENTAL FORMS OF ARSENIC (CONT.)

BIOLOGICAL SAMPLES

Type	Forms	Reference
Sea weed and epiphytes	As(III), As(V)	9, 8
	$CH_3AsO(OH)_2$	9, 8
	$(CH_3)_2AsO(OH)$	9, 8
	$(CH_3)_3As$	9, 8
Urine	As(III), As(V)	9, 8
	$CH_3AsO(OH)_2$	6
	$(CH_3)_2AsO(OH)$	6, 3
Methanobacterium Cultures	$(CH_3)_2AsH$	13, 14
Aerobic cultures (Fungi and mixed)	$(CH_3)_3As$	* 14
	$CH_3AsO(OH)_2$	*
	$(CH_3)_2AsO(OH)$	*

*Reported in this article

TABLE II. SELECTED ENVIRONMENTAL ANALYSES FOR ARSENIC

Sample	As(III)	As(V)	MAA	CA	Total
WATER					
Withlacoochee River	< 0.02	0.16	0.06	0.20	0.42 ppb
Well - near above river	< 0.02	0.27	0.11	0.30	0.68 ppb
Remote pond (Withlacoochee Forest)	< 0.02	0.32	0.12	0.62	0.06 ppb
Residential Lake A - Tampa	0.76 (sum)		0.07	0.16 +0.14 TMA	1.13 ppb
Residential Lake B - Tampa	2.74	0.41	0.11	0.32	3.58
Residential Lake C	0.89	0.49	0.22	0.15	1.75
Saline Bay Water	0.12	1.45	< 0.02	0.20	1.77
Tidal Flat	0.62	1.29	0.08	0.29	2.28
McKay Bay	0.06	0.35	0.07	1.00	1.48
BIOLOGICAL SAMPLES (wet wt.)					
Human urine (4 sample average)	1.9	3.9	1.8	15	22.6 ppb
Sargassum weed (Bermuda)	1.8	17.7	0.01	0.185	19.5 ppM
Sargassum weed (Gulf of Mexico)	0.91	4.3	0.005	0.064	5.8 ppM
Shrimp (Sargassum community)	6.4	3.0	tr	0.072	9.5 ppM

Methylarsenic compounds were found, nevertheless, in large amounts associated with the palegic Sargassum weed biocommunity on this same research cruise. Using the same analytical methods as reported by us, Carpenter and Crecelius (8) have also found methylarsenic compounds including also possibly trimethylarsine associated with Kelp plankton and shrimp tissue from the North Pacific.

Methylarsenic compounds are also associated with lakes and ponds, and are especially high where the pond is surrounded by fertilized lawns as in residential areas. Polluted seawater bays also contain the alkylarsenic compounds. As is shown in Table II, the amount of inorganic arsenic generally exceeds the amount of organic arsenic present.

Organic arsenic in biological specimens could exceed the amount of inorganic arsenic present. This was found to be the case in human urine and is also suggested by Peoples, Lakso and Lais (3) in their work with animals.

Air analyses for volatile arsenic compounds have been only recently underway and it is not possible to be completely general. Dimethylarsine and trimethylarsine have been detected in air, but only in arsenic treated areas. Methylarsenic compounds were well below 1 ng/m^3 in other ambient air samples. A few analyses of air from houses have shown <0.5 ng/m^3 methylarsenic compounds. Trimethylarsine has been detected in air over a lake containing methylarsenic compounds. Arsine itself may be produced in soil treated with arsenicals but more work is needed to confirm this.

Environmental Transformations

Much work has been done with cultures. Early evidence for biomethylation of inorganic arsenic to the methylarsenic acids and volatile forms is in the work of Challenger (12). McBride and Wolfe (13) later showed that anaerobic cultures of methanebacterium could produce what they identified as dimethylarsine. It is also possible to obtain trimethylarsine from cultures of aerobic fungi from sewage plants as reported by Cox in this symposium (14).

We have found that a pond water sample (mixed culture) having arsenite added and a nutrient media added, after a period of time will produce the alkyl-arsenic acids and trimethylarsine under aerobic conditions. In these experiments it was found that methylarsonic acid appeared first, followed by dimethylarsine and finally followed by trimethylarsine or trimethylarsine oxide. The presence of bottom sediments and anaerobic conditions were not necessary to achieve the biomethylation. Thus, methanobacterium are not the sole source of environmental biomethylation. Core samples from a lake and a pond exhibiting methylarsenic acids were taken and analyzed by section for alkyl arsenic compounds. Results indicated a maximum percentage of methylarsenic acids in the top layer but by no means in high concentrations. The absence of methylarsenic compounds below the top layer of bottom core indicates that the methylarsenic acids may remain in the aqueous phase rather than become associated with sediment. Dealkylation could also occur in the sediment. Methylarsenic acids could be carried down with natural ferric hydroxide precipitation processes.

Radiochemical techniques have been used to show that dimethylarsinic acid and methylarsonic acid (or salts) when applied to soil are at least partly lost by volatilization (15, 16).

By the use of the air analysis technique to analyze air trapped under glass (see Figure 2), a number of arsenic compounds have been found to be converted to volatile dimethylarsine and trimethylarsine in soil. Table III shows the collection of arsines under the glass jugs as a function of time of day over a two-day period. Dimethylarsinic acid treated soil gave the more striking results. Note that methylarsines are immediately observed with dimethylarsine decreasing slowly. Trimethylarsine evolution builds up slowly, although observed even in the first two-hour sampling period. A similar study was made with sodium arsenite, methylarsonic acid and phenylarsonic acid. Sodium arsenite exhibited a several-hour induction period prior to detection and a slow evolution of trimethylarsine.

Methylarsonic acid exhibited a comparatively more rapid production of trimethylarsine than arsenic(III). Soil treated with substantial amounts of phenylarsonic acid did not produce any detected arsines over the first 12 hours of the experiment. After 13 days phenylarsonic acid treated soil did evolve volatile arsines, particularly trimethylarsine. Aside from the dimethylarsinic acid experiments, trimethylarsine was the chief arsine observed in these experiments. This observation could be attributed to the pKa values and their effect on reduction of the methylarsenic acids. Dimethylarsinic acid has a pKa of 6.13 while the others have pKa values on the order of 2. Arsenic acid reduction appears to require acids in the undissociated form before reduction will proceed.

Nevertheless, biomethylation reactions of all arsenic compounds appear to proceed through dimethylarsinic acid and if this is so, why is dimethylarsine not observed together with trimethylarsine in every case? Two mechanisms could be functioning in these soil experiments. One could be a reducing process which operates to reduce free dimethylarsinic acid to the arsine; the other methylates. Biomethylation to trimethylarsine of dimethylarsinic acid would then have to proceed in a manner involving a complex or surface action of some type. Those working with laboratory culture biomethylation have mixed identifications. Challenger (12) reported trimethylarsine in aerobic cultures, McBride and Wolfe (13) reported dimethylarsine in anaerobic cultures while our work gave mixed products.

It is possible that in anaerobic conditions that reduction of the dimethylarsenic intermediate to the free arsine could be more rapid than in the presence of oxygen. Furthermore, the presence of mixed products could result from two types of biomethylation occurring in soil at the same time. Clearly further work is

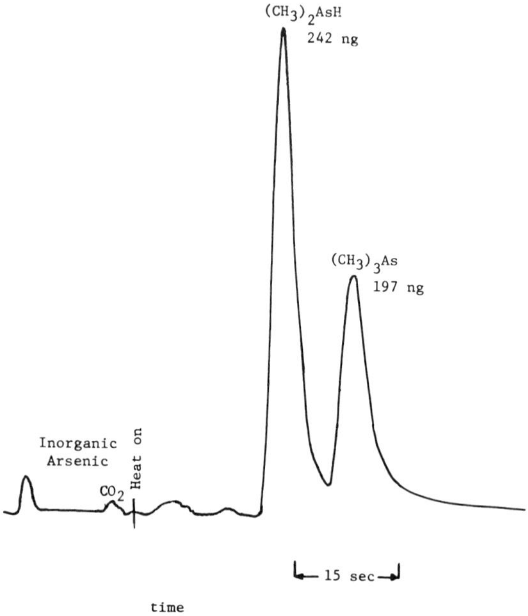

Figure 1. Air sample analysis for dimethylarsinic-treated soil. Relative emission intensity—228.8 nm As line

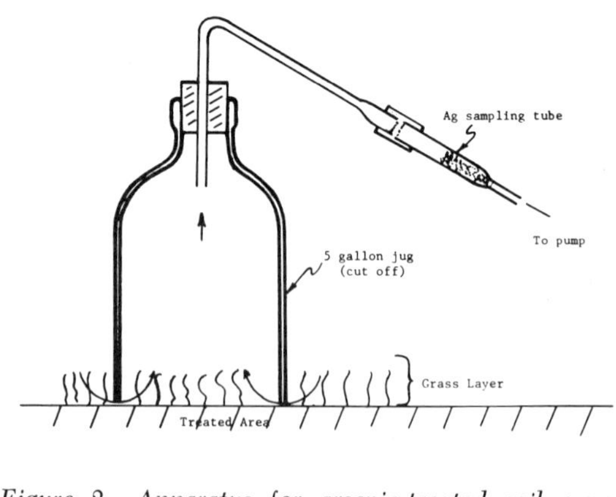

Figure 2. Apparatus for arsenic-treated soil experiments

TABLE III. EVOLUTION OF ARSINES FROM AN ARSENIC-TREATED LAWN GRASS[a]

A. Sodium Arsenite: 140 mg As/0.056 m^2						B. Methylarsonic Acid: 11.3 mg Ag/0.056 m^2	
Time Period	DMA ng.	TMA ng.	Time Period	DMA ng.	TMA ng.	Time Period	TMA ng.
5–7 PM	0	0	5–7 PM	0	1.6	5–7 PM	0
7–9 PM	0	0	7–10 PM	tr	3.7	7–9 PM	30
9–11 PM	0	0	10–12 PM	0	4.4	9–11 PM	24
11–1 AM	0	0	12–4 AM	0	7.5	11–3 AM	62
1–3 AM	0	tr	4–7 AM	0	2.1	3–7 AM	39
3–5 AM	0	1.6	7–9 AM	0	2.0	7–9 AM	HIGH[b]
5–9 AM	0	tr	9–11 AM	0	6.7	9–11 AM	460
9–1 PM	tr	14	11–1 PM	0	5.5	11–1 PM	130
1–3 PM	0	1.9	1–3 PM	5	5.5	1–5 PM	176
3–5 PM	0	3.6	3–5 PM	3	9.8	5–7 PM	70
						7–9 PM	73

[a]St. Augustine grass [b]not separated ∿500 ng [c]not well separated

TABLE III. EVOLUTION OF ARSINES FROM AN ARSENIC-TREATED LAWN GRASS[a] (Continued)

C. Dimethylarsinic Acid: 10.5 mg As/0.056 m^2

Time Period	DMA ng.	TMA ng.	Time Period	DMA ng.	TMA ng.
5-7 PM	37	tr	3-5 PM	242	197
7-9 PM	27	23	5-7 PM	58	85
9-11 PM	19	57	7-10 PM	46	266
11-1 AM	36	111	10-12 PM	5	22
1-3 AM	48	131	12-4 AM	63	241
3-5 AM	47	71	4-7 AM	48	126
5-7 AM	41	306	7-9 AM	39	77
7-9 AM	21	77	9-11 AM	179	231
9-11 AM	—— 572[c] ——		11-1 PM	500	288
11-1 PM	—— 472[c] ——		1-3 PM	84	126
1-3 PM	247	123	3-5 PM	82	97

D. Phenylarsonic Acid: 15 mg As/0.056 m^2

Initial 12 hours: no arsines detected

After 13 days - 3 hour period

Inorganic As:	2.23 ng
DMA:	1.6 ng
TMA:	11 ng

DMA - dimethylarsine

TMA - trimethylarsine

[a]St. Augustine grass [b]not separated ∿500 ng [c]not well separated

indicated.

The possibility of volatilizing arsenic out of fertilized soil was explored using a pelletized phosphate-containing commercial fertilizer. Inorganic arsenic is a known trace constituent of phosphates used in fertilizer. A 20 ft x 20 ft patch of fertilized grass was prepared using the recommended 3 lb/100 ft^2 fertilizer loading. The fertilizer was "watered in" with a lake water sprinkling system. The lake water analyzed 0.76 inorganic arsenic, 0.07 ppb methylarsonic acid, 0.16 ppb dimethylarsinic acid and 0.14 ppb trimethylarsine in solution. Air was analyzed in jugs placed in the fertilized grass patch, and over an unfertilized but watered patch of the same grass. An ambient air sampler was also set up to sample air over the fertilized patch. After a period of a few hours, small amounts of trimethylarsine were noted.

Data are given in Table IV. It would appear from this that the pelletized grass fertilizer does not cause a large increase in arsine evolution. It seems likely, in fact, that the trimethylarsine produced came from the soil biomethylation of the 0.76 ppb inorganic arsenic in solution in the lake water.

Conclusions

The presence of methylarsenic compounds in the environment is confirmed. Adequate analytical methods for their study are available. Early work indicates that commercially used arsenicals are biomethylated to volatile forms which must at least have some local and possibly global migration. Even phenylarsenic type compounds can become converted to primarily, trimethylarsine.

A number of matters can now be explored. Ambient volatile arsenic in air may be well below 1 ng/m^3 in the absence of a soil contamination with arsenic. Nevertheless, air analyses have been limited to only a few locations and much further work is indicated. Arsenic in air in places of human habitation needs study. The evolution of methylarsenic compounds out of lake water needs study. Mechanisms and rates of soil biomethylation and rates of air oxidation can now be studied.

Figure 3 summarizes the environmental chemistry of arsenic.

Acknowledgement

Much of the work reported here is part of several projects completed or underway in the laboratories of the author. This work is carried out with the assistance of D. L. Johnson, M. Tompkins and J. L. Bricker.

The support of this work by the National Science Foundation, RANN program under grant number GI-43753 is greatly appreciated.

This paper was presented in part at the SNF-RANN Second Annual Trace Contaminants Conference, Asilomar, California, August 29, 1974.

TABLE IV. EVOLUTION OF TRIMETHYLARSINE FROM A WATERED AND FERTILIZED LAWN

Time Period	[a]Lake-Watered Grass	[a]Lake-Watered and Fertilized Grass	Time Period	[b]Ambient Yard Air Analysis
5:30– 7:30 PM	0 ng	< 0.6 ng	5:30 9:30 PM	< 0.3 ng/m^3
7:30– 9:30 PM	0	0		
9:30–11:30 PM	0	0	9:30– 3:30 AM	2.1
11:30– 3:30 AM	2.9	2.7		
3:30– 7:30 AM	3.3	3.7	3:30–12:30 PM	12
7:30–11:30 AM	0.8	∿ 0.4		
11:30– 4:30 PM	1.7	3.5	12:30– 7:00 PM	∿ 0.1
4:30– 7:00 PM	0	0		
7:00–11:00 PM	---	3.4	7:00–11:00 PM	< 0.3
11:00– 3:00 AM	---	1.5	11:00– 3:00 AM	2.3 DMA, 2.5 TMA
3:00– 7:00 AM	---	∿ 0.4	3:00– 7:00 AM	∿ 0.5
7:00– 1:00 PM	---	0.6	7:00–11:00 AM	< 0.4
1:00– 5:00 PM	---	∿ 0.4	11:00– 5:30 PM	< 0.3

[a] Air trapped under glass jugs

[b] Air above fertilized patch, not trapped by a glass jug

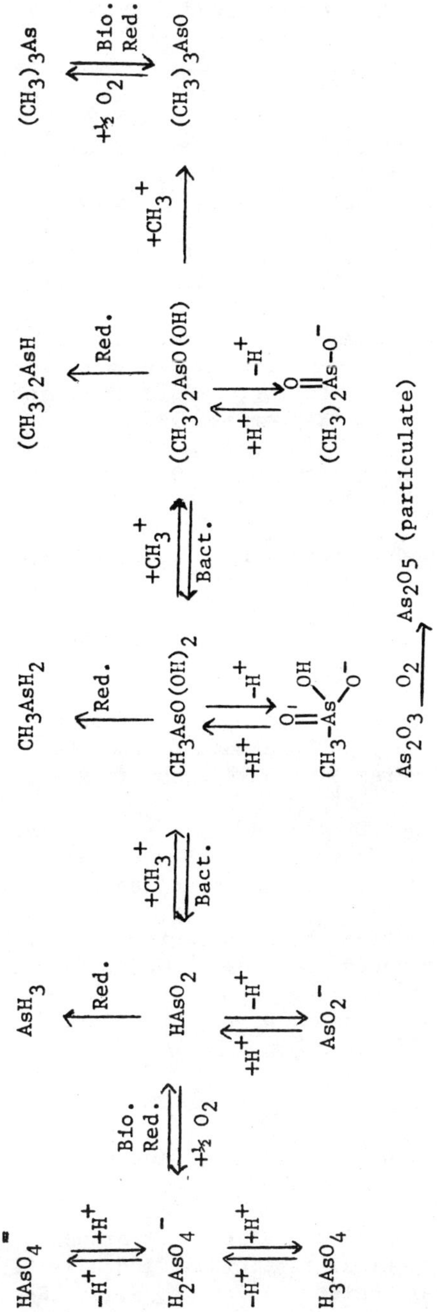

Figure 3. Summary of environmental chemistry of arsenic

Literature Cited

1. Stratton, G., and Whitehead, H. C., J. Am. Water Works Assoc. (1962), 54, 861.

2. Foreback, C. C., Ph.D. Dissertation, "Some Studies on the Detection and Determination of Mercury, Arsenic and Antimony in Gas Discharges", University of South Florida (1973).

3. Peoples, S. A., Lakso, J., and Lais, T., Proc. West. Pharmacol. Soc. (1971), 14, 178-182.

4. Talmi, Y., and Feldman, C., "The Determination of Trace Amounts of Arsenic", Symposium on Review of Arsenical Pesticides, 168th National Meeting, American Chemical Society, Atlantic City, N. J., (September, 1974).

5. Johnson, D. L., and Pilson, M. E. Q., Anal. Chim. Acta (1972), 58, 289.

6. Braman, R. S., and Foreback, C. C., Science (1973), 182, 1247-1249.

7. Braman, R. S., Justen, L. L., and Foreback, C. C., Anal. Chem. (1972), 44, 2195.

8. Crecelius, E. A., Ph.D. Dissertation, "The Geochemistry of Arsenic and Antimony in Puget Sound and Lake Washington, Washington", University of Washington (1974).

9. Johnson, D. L., and Braman, R. S., "The Speciation of Arsenic and the Content of Germanium and Mercury in Members of the Pelagic Sargassum Community", unpublished report (1974).

10. Gordon, G. E., and Zoller, W. H., Proceedings, First Annual NSF Trace Contaminants Conference, ORNL, (August 1973), p.314.

11. Braman, R. S. and Johnson, D. L., Env. Sci. and Technol. (October 1974), in press.

12. Challenger, F., Chem. Rev. (1945), 36, 315.

13. McBride, B. C., and Wolfe, R. S., Biochemistry (1971), 10, 4312.

14. Cox, D. P., "Microbial Methylation of Arsenic", Symposium on the Review of Arsenical Pesticides, 168th Nat'l. Meeting, American Chemical Society, Atlantic, N.J., (September 1974).

15. Woolson, E. A., and Kearney, P. C., Environ. Sci. Technol. (1973), 7, 47-50.

16. Isensee, A. R., Kearney, P. C., Woolson, E. A., Jones, G. E., and Williams, V. P., Environ. Sci. Techno. (1973), 7, 841-5.

17. Hiltbold, A. E., "Behavior of Organoarsenicals in Soil", This Symposium, 1974.

9

A Proposed Arsenic Cycle in an Agronomic Ecosystem

GARY R. SANDBERG and INGRID K. ALLEN

The Ansul Co., Weslaco, Texas 78596

A. Introduction

Arsenical pesticides have been widely used in agriculture for more than a century and they represent the largest single man-made input of arsenic into the environment. Copper acetoarsenite (Paris green) was first used in 1867 followed by many other inorganic arsenicals in the early 1900's. Inorganic arsenicals, such as arsenates of calcium, copper, lead and sodium, potassium and sodium arsenites, arsenic trioxide and arsenic acid were commonly used as insecticides, herbicides, algicides or desiccants.

Organic pesticides have replaced most uses for the inorganic arsenicals in the past few decades, with organic arsenicals replacing the inorganic types for herbicidal applications. The overall use patterns of arsenical pesticides have shifted from general use, broad spectrum pesticides to herbicidal uses only. Organic arsenicals marketed in agriculture today are cacodylic acid (CA) (hydroxydimethylarsine oxide), MSMA (monosodium methanearsonate) and DSMA (disodium methanearsonate).

Effects of Inorganic Arsenicals on Crop Yields and Arsenic Residues in Soil. Attention was focused on inorganic arsenical pesticides when accumulations of arsenic in soils eventually became toxic to several agricultural crops (1) (2) (3) (4) (5) (6) (7). Treatment rates for effective control of agricultural pests were usually quite large compared to pesticide applications today. In the State of Washington, the estimated peak use of lead arsenate insecticides occurred in 1943 when the average annual rate of application was 56.0 kg of elemental arsenic per ha (7). Arsenic trioxide was used as a soil sterilant to control vegetation at rates of 504.0 to 2,520.0 kg/ha (8).

The effects of these massive applications on soil residue levels were obvious. Woolson (9) reported the average arsenic levels in virgin soils at 2.5 to 10.0 ppm As and levels as high as 73 to 335 ppm As in soils with arsenical application histories. In another survey, the average arsenic level measured in 58 contaminated soils was 165 ppm As and ranged from 106 to 2,553 ppm As. Soils from Washington averaged the highest (627 ppm As), whereas soils from check plots averaged 13.0 ppm As (10).

Canadian soils contained up to 121.0 ppm As when lead arsenate insecticides were used in apple orchards (11). Turf soils treated with lead arsenate and arsenic pentoxide for grub control in 1934-37 were analyzed at 130-550 ppm As. Untreated soils were reported at 12-13 ppm As (12).

Sodium arsenite was applied directly to soil at 90, 180 and 720 kg/ha to represent long-term applications for potato vine desiccation. These treatments increased average soil arsenic levels by 20.0, 70.0 and 147.0 ppm As, respectively, the first year and 23.4, 41.4 and 96.4 ppm As three years after application (13). This use in potato fields was apparently not as detrimental to soils as insecticidal uses. A 1970 survey of Wisconsin soils that were known to have received treatments of sodium arsenite for potato vine desiccation determined that arsenic levels ranged from 2.2 to 25.7 ppm As (14).

Lead arsenate sprays for 20-25 years in Oregon increased arsenic residues in soil to 40-115 ppm As. Untreated soils contained 2.5-4.6 ppm As (15). The persistence of inorganic arsenical applications was emphasized when phytotoxic effects of 2,690 kg/ha sodium arsenite were observed 14 years after treatment (16).

It should be noted that phytotoxicity due to arsenic residues in soil are not necessarily correlated with total soil arsenic. Woolson, et al (10) found a growth reduction correlation of 0.74 when compared to total soil arsenic from 58 contaminated soils. A correlation of 0.82 was noted when plant growth was compared to arsenic fractions of Fe, Al or Ca. Soils with high reactive levels of Al were less phytotoxic than soils with low reactive Al levels; young corn was tolerant to treatments of 670 ppm As in soil that was highly reactive in Al.

Vandecaveye, et al (3) concluded that poor growth of alfalfa and barley was due to the amount of readily soluble arsenic (3.4-9.5 ppm As) in old orchard soil. Another study noted that the amount of water soluble arsenic in soil was directly related to the rate of sodium arsenate applied and inversely related to time and the amount of reactive Fe, Al or Ca in the soil. Reaction of arsenic in the water soluble form to the Fe-As form was slower than to the Al-As form, but Fe-As was less soluble (17).

Soil texture has been related to arsenic fixation (18) but both reactive Fe and Al in soils usually vary directly with clay content. Much larger rates of sodium arsenite were required to sterilize heavy soils than light textured soils (19); red soils also required higher rates of application (4). Toxicity of arsenic trioxide treatments was greater on coarse-textured than fine-textured soils (8).

Soil acidity influenced arsenic toxicity in certain soils (20) (21). Toxicity was greater in acid soils than alkaline but may be altered by As-P ratios. These results may also reflect the role of pH on the availability of reactive Fe, Al and P in soils.

Woolson (22) concluded that plant growth in soils contaminated with arsenic was affected by the amount of available phosphorus in the soil solution. When considering water soluble forms of arsenic and water soluble phosphorus, a P/As ratio of less than 4.0 significantly reduced plant growth in Lakeland loamy sand, Hagerstown silty clay loam and Christiana clay loam soils. This ratio response was valid only if the potential soluble arsenic in a given soil was large enough to be toxic in the first place. Treatments of 100 ppm As as sodium arsenate increased available arsenic levels to 27.9 ppm As in the Lakeland soil and 26.1 ppm As in the Christiana soil but growth of radishes was reduced to 23 and 93% of the controls respectively. The P/As ratios were 4.4 in the former soil and 19.2 in the latter soil.

Organic matter did not appear to have any effect on arsenic sorption (23) when sodium arsenite was added to a Superior clay loam, Waupan silty clay loam or a Plainfield sand. Arsenic sorption was increased with increasing amounts of extractable Fe_2O_3 and Al_2O_3 in soil; the greatest sorption of arsenic was observed with the Superior clay loam soil (0.5% organic C and 2.34% free iron oxides). The least arsenic sorption was noted with the Plainfield sand with 0.7% organic C and 0.34% free iron oxides.

Excessive soil moisture was found to increase phytotoxicity of As_2O_3 to Monterrey pine seedlings by 25% (24). A dose of 8,960 kg/ha As_2O_3 was required to reduce seedling survival to less than 5% in a Chenango silt loam but only 6,720 kg/ha were required under excessive moisture conditions.

Effects of Organic Arsenicals on Crop Yields and Arsenic Residues in Soil. In perspective, organic arsenical herbicides are applied at considerably lower single and annual rates than the inorganic forms and have less impact on phytotoxicity to crops or soil residues.

CA and MSMA applications contribute 9.1 and 5.5 kg respectively of elemental arsenic per hectare per year when applied at maximum recommended rates[1/].

Soil arsenic residues in the plow layer were increased an average of 3.0 and 4.5 ppm As after six annual applications of MSMA or CA respectively at recommended rates (25). Exaggerated doses of MSMA at 110 kg/ha over a six-year period increased arsenic levels in soil from 16.0 to 21.2 ppm in the 0-15 cm depth of soil. Arsenic levels in soil were elevated from 1.6 to 7.8 ppm with doses of 27 kg/ha of MSMA. None of these treatments reduced cotton yields. Soil types were Decatur silt loam, Hartsells fine sandy loam and Dothan loamy sand (26).

1/Based on three annual applications of CA at 5.6 kg ae/ha/treatment and MSMA at 4.5 kg ai/ha/treatment recommended by The Ansul Company.

Organic arsenicals are not active in the soil at normal rates of application. Data from studies to observe accumulative effects of annual CA or MSMA applications indicate that soil residues have not been phytotoxic to a wide variety of crops. Yields of soybeans, sugarbeets and wheat were not reduced after two yearly preplant treatments of CA at 8.4 kg ae/ha or MSMA at 6.7 kg ai/ha. Sugarbeets and wheat yields were tolerant to single applications of MSMA or CA as high as 67.2 or 84.0 kg/ha, respectively; soybean yields were not affected by applications of MSMA at 22.4 kg/ha or CA at 28.0 kg/ha. Soil type was a Hidalgo sandy clay loam (27).

Unpublished data from The Ansul Company concluded that corn, cotton and grain sorghum yields have not been adversely affected by six annual preplant applications of MSMA at 6.7 kg ai/ha or CA at 8.4 kg ae/ha to a Hidalgo sandy clay loam soil. Cotton yields were slightly increased with a single dose of either herbicide as high as 280 kg/ha. Corn and grain sorghum data, collected the second year but not the first year after treatment, showed that sorghum yields were not reduced by either herbicide at 280 kg/ha and that corn was tolerant to 224 kg/ha MSMA or 280 kg/ha CA. A rate of 224 kg/ha is equivalent to approximately 15 yearly applications of MSMA or CA at maximum recommended rates.

Four applications of DSMA and MSMA at rates ranging from 2.23 to 8.95 kg/ha per treatment were made for four successive years on a Chesterfield sandy loam. Yields of cotton, soybeans, sorghum-sudan hybrid, corn, oats, vetch and crimson clover were not reduced by any of the treatments. Average soil arsenic residues varied from 10.4 to 26.7 ppm above background in soils treated with the low and high rates of MSMA; DSMA applications increased levels 8.4 to 17.3 ppm As. The total depth of soil sampled was 30 cm. About 90% of the soil arsenic was associated with Al in the clay fraction while phosphorus was identified in the Fe and organic matter fractions (28).

DSMA applications at 134.4 kg/ha were not phytotoxic to winter or summer crops grown for two years after treatments in Arizona. Winter crops were barley and safflower; summer crops were cotton and sorghum. Soil texture was 42% sand, 37% silt and 21% clay (29).

A greenhouse study concluded that DSMA treatments of 50 ppm inhibited rice growth by 75%, soybeans by 39%, oats by 12%, corn and cotton by 10% and wheat by 0%, when incorporated into a Bosket silt loam. Incorporations of 100 ppm DSMA were not phytotoxic to rice in Dubbs silt loam or Sharkey clay. The 50 and 100 ppm doses were calculated for soil on an oven dried basis so the equivalent rates were greater than 112 and 224 kg/ha, respectively, based on a 6 in. furrow slice of soil. Incorporations of DSMA at 120 ppm (268.8 kg/ha) in the Bosket silt loam reduced cotton growth initially but did not inhibit growth 32 weeks later. Additions of phosphorus increased arsenic toxicity in both of the loam soils but not the clay soil (30). Apparently the amount of phosphorus added was sufficient to displace DSMA or its breakdown product (probably arsenate) to increase the amount of available arsenic in the soil solution. Woolson (personal conversation) found that low levels of phosphorus added to an arsenic toxic soil will displace arsenic from soil particles to increase toxicity to plants but that large applications of phosphorus will compete with arsenic at the root surface and actually decrease toxicity. Also, high phosphorus additions may increase arsenic uptake by plants without causing toxic symptoms.

Other researchers (31) found that the amount of DSMA adsorbed in Norfolk loamy sand, Augusta silt loam, Decatur clay loam and Vaiden clay soils was 7.7, 34.3, 27.2, and 50.2% of the equilibrium solutions added for each soil, respectively. These values corresponded to the clay fractions of each soil type.

Lower application rates of arsenic and lack of soil activity associated with normal organic arsenical applications offer a distinct advantage over inorganic forms; however, the additional importance and uses of these products in agriculture have reemphasized the need to understand the behavior of arsenicals in the environment and the potential hazards of their continued use.

B. Background Arsenic in Nature

To understand conditions in which arsenicals could accumulate to undesirable levels in nature, it is necessary to examine the distribution of arsenic in nature and the mechanisms which affect the arsenic cycle.

In terms of abundance of elements, arsenic ranks 18th in the universe, 20th on the earth's crust, 14th in sea water and 12th in the human body (32).

Arsenic levels of 13 ppm in shales but only 1.8 ppm in igneous rock and 1.0 ppm in sandstone and limestone (33) suggests the influence of biological systems on the accumulation of arsenic during early geological times. Relatively high levels of 140 ppm As in coal (34) and 42 ppm As in unweathered shale (12) also support the role of primordial plant life in arsenic distribution.

Over sixty soil types which were believed to be free of arsenical applications were surveyed by Williams and Whetstone in 1940 (12). About 30% of the samples contained less than 5 ppm As, 50% contained 5-10 ppm As and about 20% more than 10 ppm As. There was no uniformity in the distribution of arsenic within the profiles and no relationship between the climatic conditions or geographical formations from which the soils were developed. In general, the arsenic level in sandy soils and soils with a high silica : sesquioxide ratio was relatively low, and that of soils of the semi-humid and arid regions was higher. The soil with the lowest level of arsenic observed was a Brassua sandy loam (New Hampshire) at 0.1 ppm; the highest level was 42 ppm As in a Canyon clay loam (Kansas).

Unpublished data from The Ansul Company noted that the average background arsenic levels in soil ranged from 3.37 ppm As in Georgia to 11.1 ppm As in Texas. California, Arizona and Mississippi soils averaged 5.42, 5.06 and 6.16 ppm As respectively. The variability within a sampling site was as much as five to tenfold.

Detectable levels of arsenic have often been reported in freshwater systems and were not attributed to sources of contamination. Durum, et al (35) surveyed freshwater systems in the U.S. and found that 76% of 726 samples were below limits of detection (0.01 ppm As); the remaining 24% of the samples averaged 0.01 ppm As but ranged from 0.01 to 0.14 ppm As. Kopp and Kroner (36) noted that 94% of 1,577 samples collected in another survey were below 0.10 ppm As. The average As level was 0.06 ppm and ranged from 0.01 to 0.34 ppm As. The Lake Erie watershed contained the highest level (0.34 ppm As). No detectable levels were found in the California watershed. Lis and Hopke (37) concluded that the average arsenic level in a New York lake was 0.01 and ranged from 0.01 to 0.04 ppm As.

Schroeder and Balassa (38) reported arsenic in many common foods from commercial sources. Fish and seafood averaged 4.64 ppm As with shrimp shells being the highest component (15.3 ppm As). Meats ranged from below limits of detection (<0.1 ppm As) to 1.4 ppm As in pork liver and averaged 0.49 ppm As for all meats. Vegetables and grain averaged 0.41 ppm As; corn meal was highest with 0.78 ppm As.

A survey of 28 U.S. cities during 1969-70 concluded that arsenic levels in dairy products were less than 0.1 ppm As but ranged from 0.1 to 2.6 ppm As in the meat, fish and poultry food class (39). Virtually all living organisms are exposed to arsenic. The role of arsenic as an essential element has been suggested (40) (41) (42) (43) but not proven.

C. Review of Arsenic Cycles and Related Ecological Studies

Studies to observe the movement of arsenic between given phases in an ecosystem have been complicated by the absence of a stable arsenic isotope and large quantities of analytical data to establish reliable background levels of arsenic. Natural arsenic levels are characterized by large variations within samples and between sampling sites.

Frost (43) proposed a closed arsenic cycle in nature that continually transferred arsenic between water, soil, air and animals; volatile arsines were characterized as the form of arsenic in air and contributed to soil and water arsenic levels. Methylarsines are degradation products of organic arsenicals and play a very significant role in the arsenic cycle. More discussion on methylarsines is presented in the proposed cycle.

Allaway (44) described the environmental pathways of trace elements to include arsenic, but did not estimate the input by agriculture. Annual uptake of arsenic by vegetation was a major transfer mechanism in the arsenic cycle.

Isensee, et al (45) studied the distribution of CA and dimethylarsines (DMA) in an aquatic ecosystem. Neither chemical showed a high potential for biomagnification in algae, *Daphnia*, snails or fish. Treatments at 0.1, 1.0 and 10.0 ppm CA or DMA were not toxic to the organisms.

A thesis by Macklin (46) observed the effects of CA tree injections[2/] on a forest ecosystem. A large amount of the arsenic applied was transferred to the forest floor with leaf drop (1,050 ppm As), but subsequent movement of arsenic from the litter to soil was relatively small; soil arsenic levels were increased from 15.4 ppm As (before treatment) to 19.2 ppm As 16 weeks after treatment. Final concentration of arsenic in the litter was 96.0 ppm which suggests that a loss other than leaching was occurring, probably reduction to volatile methylarsines.

Greaves (47) discussed some of the factors affecting the distribution of organic arsenical silvicides in the forest ecosystem. It was concluded that the movement of arsenic with litter fall and decomposition of woody tissue could be controlled with timing and density of treatments. In the forest ecosystem, various parts of the tree could be "sinks" in the model system which would minimize exposure to non-target organisms.

Norris (48) found that arsenic was mobile in the forest floor litter after treatments of CA or MSMA for forest thinning operations, but that once the arsenic reached the soil it was rapidly fixed. The soil represented a large "sink" in this study but it was suggested that redistribution of arsenic from the litter may occur with methylarsines.

Malone (49) (50) studied the effects of CA on a fescue meadow. The responses of microorganisms, plant biomass and community structure were essentially equal to those of the controls.

Other ecological studies have been conducted with organic arsenicals to assess effects on community structures. In his thesis, Edwards (51) found that 30 sprayings of MSMA at total doses of 3,000, 30,000 and 300,000 ppm MSMA over a three-month period had relatively little effect on a salt marsh ecosystem. The highest rate was toxic to *Spartina alterniflora*, the major vegetation in southeastern marshes, and *Littorina irrorata*, a salt marsh mollusc, but not the lower rates. Arsenic levels in two mollusc species were high initially after treatment but decreased appreciably within one month after treatment. Spraying or flooding methods of MSMA applications produced equal results on *Spartina*.

Hunter and Young (52) surveyed the effects of military defoliants on vegetative succession in Florida. CA was applied at total rates of 12.3 and 59.4 kg ae/ha over an 8 year period; 2,4-D, 2,4,5-T and picloram were also used in the 240 acre site. Soils within the treated area were Lakeland, Chipley and Rutledge

[2/] A CA formulation is marketed commercially as a silvicide.

sands and ranged from poor to moderate in drainage. Nine months after the last application, all but two of the 74 dicotyledonous species collected in untreated areas were found in the treated site. Vegetative composition and successional patterns were not affected by defoliant treatments or application patterns but were apparently influenced by soil type and previous mechanical disturbances.

In a similar study (53) Pate, et al determined that species diversity among mammals, birds, reptiles and amphibians was large in treated and untreated areas. The observed differences in species found could be accounted for in habitat preferences, especially vegetative cover, but not chemical treatments.

D. A Proposed Arsenic Cycle in an Agronomic Ecosystem

The cycle proposed in this paper considered plant, soil, air and water, with special emphasis on the soil since that is assumed to be the "sink" for the system. The inputs of arsenic into the model were primarily CA and MSMA applications, but additions with fertilizers, irrigation water, and oxidation of arsines were also considered. Industrial and municipal contributions were recognized but the amount of contamination from this source was not estimated.

The most important transfer mechanisms in the model, or for any arsenic cycle, include arsenical oxidation and reduction reactions, soil erosion, plant uptake and translocation and harvesting of crops for food, feed and fiber.

The transfer rates used in this model are often quite general and based on available data at this time. Additional studies are required to identify the effects of specific factors on the arsenic cycle; some factors suspected are soil type, soil chemistry, organic matter, climate, agronomic cultural practices, crops grown, arsenical use history and others. This cycle is based primarily on field crops grown in the Lower Rio Grande Valley of Texas but the basic concepts are applicable to other types of agriculture such as forest, pasture, orchard or even landscape applications.

It should be noted that the arsenic cycle will continue to operate without an agricultural input but at a greatly reduced rate. In the absence of any arsenic inputs, the soil depletion time was estimated to be 100 years due to plant uptake and mining alone (44).

A diagram of the proposed cycle is presented in Figure 1. Each pathway is numbered and discussed individually.

Transfer 1: Applications of MSMA or CA Herbicides. The primary inputs of arsenic into this system are post-emergent applications of organic arsenical herbicides. The number and rate of applications vary with chemical used and with crops treated. The maximum recommended yearly doses of either chemical were selected

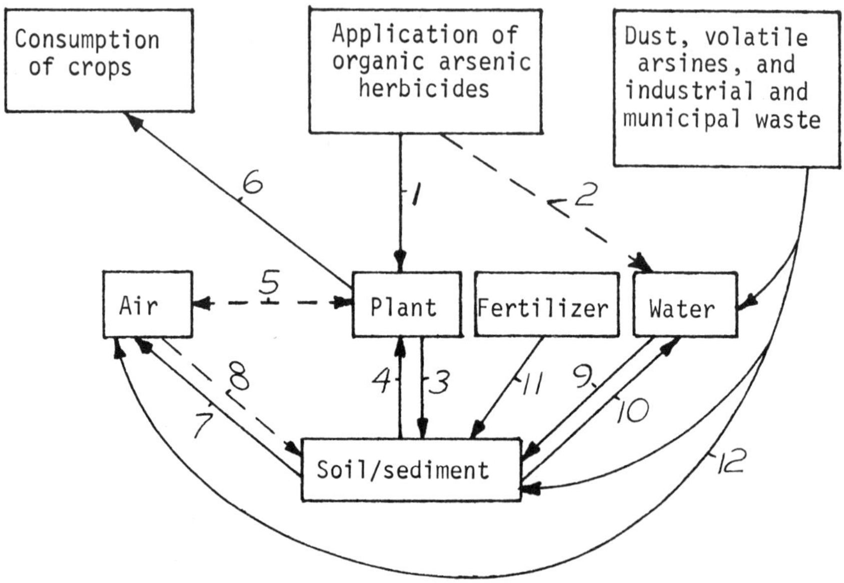

Figure 1. A proposed model for the arsenic cycle in an agronomic ecosystem

for this model to demonstrate conditions which most favor a build up of arsenic in the system.

MSMA or CA herbicides are applied to certain orchard crops up to three times within a season and at rates not to exceed 4.5 kg ai MSMA/ha or 5.6 kg ae CA/ha per application. The greatest possible seasonal input is 13.5 kg/ha MSMA or 16.8 kg/ha CA. Normally the rates required for most agricultural crops are not this large, especially for weed control in field crops.

MSMA and CA contain 40.7 and 54.3% elemental arsenic respectively. The greatest input of arsenic/ha/yr is 5,495 gm (about 2.4 ppm) from MSMA or 9,122 gm (about 4.1 ppm) from CA.

Transfer 2: Contamination of Water with Spray Drift or Volatilization. Pentavalent organic arsenicals are not volatile and direct spray contact or drift can be avoided if routine application precautions are observed. In the event of a misapplication, MSMA and CA are completely water soluble and are rapidly dispersed in water. Eventually MSMA or CA are oxidized to the arsenate form and fixed on sediment particles or are reduced to methyl arsines which migrate out of the treated area.

Experimental data from applications of MSMA to irrigation ditchbanks showed that arsenic residue levels in water from treated canals reached a maximum of 0.16 to 0.86 ppm As when the spray solution directly contacted the water surface. These levels dispersed to below 0.05 ppm As (U.S. Public Health Service drinking water standards) within 80 to 100 minutes after application. The higher residue levels were noted in small irrigation laterals with relatively low flow rates (54).

The same study found that 28 and 7% of the MSMA added to nonfiltered and filtered water samples respectively could not be recovered after 36 days. This loss may be due to arsenic fixation on soil sediment particles and/or methylarsine movement out of the samples.

Isensee, et al (45) found that 92.0% of the CA and 61.2% of dimethylarsine added to an aquatic ecosystem was recovered after 24 hours. The 8% loss was attributed to initial adsorption to aquatic organisms.

It was assumed that the spray drift transfer was negligible in this model.

Transfer 3: Arsenic Movement from Treated Vegetation to Soil. The movement of arsenicals to soil from treated foliage takes place with spray runoff during application, direct spray contact with the soil surface and precipitation after treatment (rain, dew, sprinkler irrigation, etc). These transfers occur rapidly and do not allow chemical reactions to other forms of arsenicals.

Other plant-to-soil movements are found with the death and decay of treated foliage and with translocation of MSMA and DSMA to the stems and roots; CA does not translocate and is strictly a contact herbicide. It has been reported that biochemical complexing of organic arsenicals takes place in plant tissues; also, these transfers take place at relatively slow rates.

Translocation of MSMA and DSMA in susceptible weeds and tolerant crops has been investigated by several researchers. Johnsongrass and nutsedge applications were most commonly reported but evidence of MSMA and DSMA translocation in cotton and bermudagrass has also been cited.

Duble, et al (55) found that less than 15% of the DSMA or amine methylarsonate (AMA) applied to purple nutsedge moved out of the treated shoots. It was also noted that DSMA was not readily broken down in the plant but that a DSMA complex may have been formed in the plant.

Another study (56) concluded that arsenic was translocated in purple nutsedge tubers separated by at least four tubers from the treated shoot. Arsenic content was usually higher in the tuber at the opposite end of the chain and in tubers with actively growing points. Arsenic content in the tubers was not correlated with their ability to produce new shoots.

Wills (57) found that 12.9% of the ^{14}C-MSMA applied to the foliage of purple nutsedge was absorbed into the leaf and that 11.7% of the amount applied was translocated to the tubers.

Keeley and Thullen (58) treated yellow nutsedge with 3.36 kg/ha MSMA or DSMA and found that treated tubers contained 4 to 33 ppm As while untreated tubers were found to have only 1 ppm As. Small tubers contained more arsenic (23 to 33 ppm) than large tubers (4 to 12 ppm); vitality was reduced in small tubers but not in large tubers.

In another study (59) Keeley and Thullen found that ^{14}C-MSMA and ^{14}C-DSMA treatments caused greater radioactivity in yellow nutsedge than in purple nutsedge. The differential penetration of the leaves probably contributed to poor control of purple nutsedge. Larger amounts of radioactivity in the plant were generally noted with MSMA treatments than with DSMA treatments and at warm temperatures. Approximately 85% or more of the radioactivity remained in the plant but some activity was found in daughter plants. It was suggested that nutsedge did not metabolize ^{14}C-MSMA or ^{14}C-DSMA after 72 hours of exposure.

Nutsedge treated with repeated applications of amine methylarsonate (AMA) increased starch hydrolysis in tubers but had little effect on fat and protein content. AMA-treated tubers respired at lower rates than untreated tubers (60).

Cooley (61) found that MSMA foliar applications at 2.24 kg/ha to woolly leaf bursage caused severe cell damage to root tissue. It was concluded that MSMA was translocated to the roots in herbicidal quantities since regrowth or initiation of new growth was not observed.

Sckerl and Frans (62) showed that methanearsonic acid (^{14}C-MAA) applied to the foliage of johnsongrass was translocated to the root tips and terminal leaf within 8 hours. It was suggested that MAA may complex with either a sugar or organic acid or both. MAA was translocated from treated cotton stems to all parts of the plant but moved very little from treated foliage to the terminal leaves, stems, lower leaves and petioles.

Ehman (63) reported arsenic translocation to cottonseed to levels as high as 40 ppm As if topical applications are made at the closed-boll stage of growth.

Approximately 25% of the DSMA applied to coastal bermudagrass foliage was translocated to the roots and rhizomes within 5 days. Some ^{14}C DSMA was found in respiratory CO_2 but most of the C-As bonds were intact (64).

In studies using ^{14}C-MSMA, ^{14}C-DSMA and ^{14}C-MAA, Keeley and Thullen (65) found little translocation from cotyledons to developing leaves of cotton seedlings.

It was assumed that 100% of the arsenic applied as MSMA or CA was transferred to the soil in this model. An additional input via this mechanism was the return of crop residues to the soil after harvest; this rate was calculated by subtracting the amount of arsenic transferred with harvest from the total uptake value for this model; 15.8 gms As/ha/yr (harvest loss) was subtracted from 23.7 gms As/ha/yr (total uptake) to arrive at 7.9 gms As/ha being returned to the soil as crop residues each year. More detailed explanations of these figures are presented in each transfer discussion.

Transfer 4: Uptake of Arsenic by Vegetation. Uptake of arsenic by plants is a continuous process and will occur in the absence of arsenical pesticide applications.

The presence of arsenic in vegetation growing in virgin forest soils was observed by Schroeder and Balassa (38). Various fern species contained from 0.0 to 0.73 ppm As and pine needles were analyzed at 0.32 ppm As; the forest soil contained 3.64 ppm As. Williams and Whetstone (12) found that native vegetation on natural soil did not exceed 10 ppm As.

Liebig (41) reviewed arsenic residue levels reported in several studies on almost 40 agricultural crops. This summary established low, intermediate or high ranges of arsenic levels observed for tissue analysis. Vegetables and other plants contained from 0 to 10 ppm As and usually had higher levels of arsenic when grown in soil contaminated with inorganic arsenical insecticides. It was concluded that arsenic does not normally accumulate to any extent in the above-ground parts, so tissue analysis of the tops of plants is a very poor indication of arsenic toxicity. Arsenic does accumulate in larger amounts in or on roots and causes plasmolysis or rotting of roots in solution cultures (42). Apparently the effect of arsenic toxicity retards plant growth before extensive amounts of arsenic are absorbed and translocated to the top.

Data collected by The Ansul Company (27) concluded that two annual preplant applications of MSMA or CA at rates of 6.7 or 8.4

kg/ha/yr respectively did not significantly increase arsenic residue levels above background in corn seed and fodder, cottonseed, sorghum seed and fodder, wheat, sugarbeets or soybean fodder. Arsenic residue in soybean seed was not affected by MSMA treatments but was significantly increased (0.40 ppm As) after the second annual treatment of CA.

In the same report (27), single preplant applications of MSMA or CA were used to approximate the effects of 10 years of normal applications. MSMA at 22.4 and 67.2 kg/ha did not increase arsenic residues in any of the crops mentioned above after the first or second year after treatment except soybeans; the high rate of MSMA increased residues after the first year (1.15 ppm As) and both rates significantly increased residues after the second year (0.36 and 0.58 ppm As). A single preplant treatment of CA at 28.0 kg/ha did not increase arsenic residues in any of the crops after the first year but did significantly increase residues in soybean seed (0.46 ppm As) and sugarbeets (0.63 ppm As) two years after treatment. The high rate of CA (84.0 kg/ha) applied as one preplant treatment increased arsenic residues in corn seed (0.16 ppm As), corn fodder (3.53 ppm As), cottonseed (0.32 ppm As), soybean seed (1.60 ppm As) and wheat (0.38 ppm As) after the first year; after the second year, residue levels continued to be significantly greater in corn fodder (0.89 ppm As), soybean seed (0.55 ppm As), sugarbeets (0.85 ppm As) and wheat (0.21 ppm As). It is of interest to note that crop yields were not reduced with any of these treatments except soybeans treated with 67.2 kg/ha MSMA and 84.0 kg/ha CA the first year but not the second year after treatment.

As with arsenic toxicity, uptake of arsenic by plants is not necessarily correlated with the total amount of arsenic in the soil. Woolson (22) found that arsenic uptake by six vegetable crops from soils treated with sodium arsenate was generally correlated with the amount of available soil arsenic required to reduce crop growth by 50% (GR_{50}). Residues were low in crops where the edible portion of the crop was not the root or the entire plant. A P/As ratio has been implicated in arsenic toxicity (22) and probably has the same effect upon arsenic uptake. Woolson feels that high phosphorus additions to arsenic-contaminated soils may increase arsenic uptake by plants without causing toxic symptoms.

Schroeder and Balassa (38) found that arsenic levels in vegetables and grain grown in soil never sprayed with arsenicals were essentially equal to those grown in soil treated with heavy applications of phosphate fertilizer with 4.75 ppm As. The average residues in control and treated grains were 0.93 and 0.47 ppm As respectively; vegetables were 0.25 and 0.28 ppm As.

Johnson and Hiltbold (28) found unusually high arsenic residues (1.61 to 5.20 ppm As) in several agricultural crops when grown on soils that had received previous DSMA or MSMA treatments for four successive years. The highest total dose of elemental arsenic applied during this period was 66 kg As/ha with MSMA treatments and 58 kg As/ha with DSMA treatments. A light phosphorus application was made before planting but crop yields were not affected. It was not clear what background levels were observed in untreated crops.

The variability in arsenic uptake with different crops and soil conditions has been emphasized. Since this model is an attempt to illustrate general transfers in their perspective, no attempt was made to pick one type of crop for this transfer. To arrive at a general value, data from Allaway (44) was used to calculate that 25.5 x 10^9 tons of biomass were produced each year from 12 x 10^9 acres of agricultural land. If the total arsenic concentration in plants was assumed to be 5.0 ppm As, a transfer of 23.7 gms As/ha/yr was calculated. The 5.0 ppm value for plants is probably low but was allowed to present conditions which favor a soil build-up of arsenic.

Transfer 5: Reduction of Organic Arsenicals to Volatile Forms of Arsenic (Plant-Air). Although not supported with data, it is suspected that the reduction of organic arsenicals to volatile methylarsines may take place while the dead vegetation is still standing. This reduction reaction is further discussed in the section on Transfer 7.

The form of arsenic remaining in plants was determined to be an MSMA or DSMA complex (see discussion for Transfer 3) and CA may also be complexed within the cellular tissue. The role of microorganisms in the degradation of plant tissue is not questioned and supports this loss mechanism.

Malone (50) found that treatments of CA to a fescue meadow actually increased the rate of litter decomposition from standing dead vegetation. He concluded in a similar study (49) that CA temporarily stimulated some soil bacteria and decreased the number of fungi but increased the overall decomposition process.

For purposes of modeling this system, it was assumed that this transfer was negligible. It is possible that this loss is sufficient to further decrease the potential build-up of arsenic in soil.

Transfer 6: Arsenic Losses with Crop Harvest. The portion of an agricultural crop that is harvested for food, feed or fiber represents a loss of arsenic from the agronomic ecosystem. It has been noted that arsenic residues are generally lower in seed and fruit crops than in root crops or crops harvested for the entire plant; the amount of biomass removed also follows the same pattern.

In the following summary, yield data were taken from the U.S.D.A. 1972 Annual Summary CrPr 2-1(73). Arsenic residue data were collected by The Ansul Company.

To calculate a general transfer value for this model, all of the crop losses were averaged and a value of 15.8 gms As/ha/yr was

Crop	Annual Yield (lbs/A)	Ave. As Level (ppm)	Est. As Loss (gms/ha/yr)
Alfalfa hay	5,600	0.29	1.80
Barley	2,093	0.13	0.45
Bermudagrass	4,000	1.23	5.51
Corn seed	5,040	0.10	0.56
Corn fodder	24,000	0.22	5.96
Cotton lint	450	1.27	0.67
Cottonseed	680	0.06	0.11
Oats	1,600	0.80	1.46
Sorghum seed	3,080	0.17	0.56
Sorghum fodder	20,000	2.15	48.37
Soybeans	1,500	0.12	0.22
Sugar beets	40,000	2.83	123.35
Sugarcane	84,000	0.30	28.35
Wheat	2,700	0.07	0.22

the amount lost. Trees (forest and orchard) are actually large temporary "sinks" for arsenic until harvested for lumber or taken out of cultivation. Therefore the amount of arsenic lost would be much greater than the values for field crops.

Transfer 7. Reduction of Arsenicals to Volatile Forms of Arsenic (Soil-Air). Reduction of arsenicals to volatile forms of arsenic has been recognized for many years, but the role of this reaction in the arsenic cycle is just being realized. It has previously been concluded that once an arsenical came into contact with soil it was permanently bound at that site. It now appears that the reduction of organic and, to a lesser extent, inorganic arsenicals may be a significant loss mechanism from the soil which redistributes arsenicals within the environment.

Arsine gas was discovered in 1775 by Scheele and described as a poisonous gas with a persistent garlic-like odor. Vallee, et al (66) reviewed the history and toxicology of arsines. Arsine was liberated when hydrogen was generated in the presence of arsenic but also resulted from the reduction of arsenous or arsenic acid, from electrolysis of arsenous solutions, or from the action of water or dilute acid on metallic arsenides.

The review (66) pointed out that most of the accidents involving arsines were reported in chemistry laboratories or in industry. But, in 1839, Gmelin found that a garlic odor, common in many damp European homes, was probably due to arsines that were generated from arsenic used as a coloring pigment in wallpaper and carpets. Apparently the damp and moldy environment was conducive to the reductive process.

In 1891, Gosio (67) reported the generation of arsenical gas from the action of the mold, *Penicillium brevicaule*, on potato mash containing arsenious oxide. He was not able to characterize the gas but later work suggests that it was probably dimethyl or trimethylarsine. Gosio later reported that *Aspergillus glaucus*,

A. virens, *Mucor mucedo* and *M. ramosus* also produced this gas (Gosio-Gas); the exact nature of this gas was debated but was characterized as being pungent and with a garlic odor.

Challenger, et al (68) cultivated four strains of *Penicillium brevicaule* on bread crumbs containing arsenious oxide to get trimethylarsine. Later work (69) indicated that certain fungi were responsible for the biological methylation of arsenic, but negative results were noted with bacteria. In contrast, McBride and Wolfe (70) reported the formation of dimethylarsine with *Methanosarcina* and *Methanobacterium* species of bacteria.

Smith and Cameron (71) studied the use of the mold *Scopulariopsis brevicaulis* (previously *Penicillium brevicaule*) to detect arsenic in many kinds of food. If arsenic was present in amounts of 0.001 mg or less, a characteristic garlic odor was detected in two hours; larger amounts of arsenic added (as As_2O_3) to sand produced garlic odors within 20 minutes.

Zussman, et al (72) noted that arsine generation was reported for several species of saprophytic fungi but not for dermatophytic species. In their work *Trichophyton rubrum*, the latter type, produced a strong garlic odor from cultures with arsenate but not with arsenite. It was not established if arsine or a methylated form of arsenic was produced. The gas was generated at concentrations of 0.006 to 0.0015 M arsenate; fungal growth was inhibited at higher concentrations and little odor was detected at lower concentrations.

Merrill and French (73) found that *Lenzites trabea* and *L. saepiaria*, brown-rotting fungi, liberated a strong garlic odor when grown on wood treated with an arsenical preservative or As_2O_3. Sixty-three other species of wood-rotting fungi did not cause a garlic odor when grown on a medium containing As_2O_3. All attempts to identify the gas were unsuccessful.

In a report by Thom and Raper (67), 10 strains or species of *Scopulariopsis* (or *P. brevicaule*) were active arsine gas producers and 14 strains of *Aspergillus sydowi* were positive; both of these groups are common in soil. Many of the organisms tested were not found to be gas producers at 0.1 and 0.15% arsenic (as arsenious oxide) but may be found to be positive at lower concentrations. A Durham sandy loam soil was surveyed to identify the mold species present. This soil contained enough arsenic from inorganic arsenical applications to be toxic to certain crops. About two species of *Fusarium* and two or three unidentified sterile masses were the most abundant molds. *Actinomyces* and bacteria colonies were noted and eventually *Myamoeba* and plasmodium organisms developed. No infusoria or nematodes were seen. Of these organisms surveyed, all but *Aspergillus ustus*, a Penicilla, and some of the sterile forms were active gas producers. It was concluded from this study that accumulations of arsenic in the soil may be expected to occur only when massive amounts are used or under special conditions unfavorable to the development of a varied microflora.

McBride and Wolfe (74) presented a metabolic pathway using

Methanobacterium cells or cell extracts to reduce and methylate arsenate under anaerobic conditions. In the pathway, arsenate is first reduced to arsenite which is methylated to form methylarsonic acid. Methylarsonic acid (MAA) is then reduced and methylated to form dimethylarsinic acid (cacodylic acid or CA); CA is further reduced to dimethylarsine gas.

Braman and Foreback (75) developed procedures to analyze for arsenate, arsenite, methylarsonic acid (MAA) and dimethylarsinic acid (CA). All four forms of arsenic were found in various samples of natural waters; the latter two were also detected in bird shells, seashells, and human urine. It was probable that all of these forms of arsenic occurred naturally and were representative of the biological reduction levels that are part of an arsenic cycle. CA was a major and ubiquitous form of arsenic involved in biological systems. CA was usually found in larger amounts than MAA which indicates the sequence of CA in the arsenic methylation sequence. It was noted that CA is approximately 25 times less toxic than the arsenate ion which suggests that the methylation reactions are a detoxifying process. The authors did point out that arsenic inputs may eventually result in increased levels of the methylated arsenicals in water and air due to bacterial mobilization.

Three factors that affect the generation of methylarsines are species of organism, the concentration of arsenic, and the form of arsenic. Cox and Alexander (76) found that pH may also influence the amount of trimethylarsine generated. After incubation of about one month, garlic odors were detected in cultures with dimethylarsenic acid (CA) at pH 4, 5 and 7 and in cultures with methylarsonic acid (MAA) at pH 5. Sodium arsenate was reduced at pH 4. Trimethylarsine was generated in cultures with three sewer fungi species, Candida humicola, Gliocladium rosenum and Penicillium spp., in the presence of CA and MAA. Reduction of sodium arsenite or sodium arsenate was not observed with G. roseum or Penicillium but did take place at low rates with C. humicola. Acid conditions may have enhanced arsenic reduction in this study.

Newton and Greaves (48) studied the effects of levels of CA or MAA and of glucose levels on the loss of arsenic through volatilization. The losses were probably due to an interaction between the two variables; substantial losses were noted at all glucose levels while the amount of arsenic lost after 28-31 days was generally greater with the high concentrations of CA or MAA. Variations in the losses may have been due to a substrate-organism specificity. All losses were recorded at temperatures below 70°F.

Woolson and Kearney (77) found that arsenic losses from soil were influenced by soil type, concentration of CA, and moisture levels. Losses attributed to methylarsines were greater at the highest rate of CA applied and under anaerobic conditions. Methylarsine generation was highest in the Lakeland loamy sand and about equal in the Hagerstown silty clay loam and Christiana clay soils. The average losses were 35 and 61% under aerobic and anaerobic conditions respectively during a 24-week period. It was concluded that the gas may have been dimethylarsine which is extremely unstable and could be oxidized back to CA which is probably returned to plants or soil.

The Ansul Company (25) concluded that as much as 17% of the MSMA and 28% of the CA applied to soil over a six-year period could not be accounted for in the soil profile. Total doses for the period were 40.2 kg MSMA/ha and 50.4 kg CA/ha. It was concluded that the losses were greater from plots treated with high rates than those treated with low rates. Arsenic not recovered was attributed to migration of methylarsines out of the test area.

It should be pointed out that while arsines are known to be highly toxic to humans, these gases are also volatile and very unstable at high concentrations. Vallee, et al (66) stated that simple precautions could have prevented all of the reported fatalities in industry or in chemistry laboratories. Usually workers have failed to realize the risk of moisture or acid containing impure metal, especially in closed areas. In humans, arsine concentrations of 3-10 ppm may cause toxic symptoms in several hours, 10-60 ppm may be dangerous in 30-60 minutes, and 250 ppm may be lethal in 30 minutes. The maximum average safe concentration for the exposure of workers to arsine is 0.05 ppm as recommended by the American Conference of Governmental Industrial Hygienists. It is reasonable to conclude that the levels of volatile arsenicals generated in biological systems under normal conditions are within the 0.05 ppm concentration limit.

For this model, it was assumed that 17-35% of the total input of arsenic to the soil was lost as methylarsines; this included inputs from crop residues, irrigations, and fertilizers. The loss calculated was 1,638-3,282 gm As/ha/yr for CA applications and 1,018-2,007 gm As/ha/yr for MSMA applications.

Transfer 8: Oxidation of Methylarsines. Methylarsines are very unstable in air at concentrations above .05-.10 ppm and are rapidly oxidized to less reduced forms of arsenic. Problems associated with residue analysis for the methylarsines are evidence of their instability. The pathway proposed by McBride and Wolfe (74) would be reversed so that the oxidation pathway would be a methylarsine-CA-arsenate sequence. Conclusions by Braman and Foreback (75) and Woolson and Kearney (77) also support this oxidation pathway.

Arsenic losses reported in recovery studies by Woolson and Kearney (77) and The Ansul Company (25) indicate that at low concentrations methylarsines were stable enough to allow migration out of treated areas. The presence of CA and MAA in bird shells, seashells and water in areas remote from agriculture was evidence of migration of naturally occurring arsenic.

Other studies have reported losses in which all of the arsenic applied to soil was not accounted for. Epps and Sturgis (78) found that 21.6, 3.8 and 1.3% of the initial arsenic levels in a flooded Crowley silt loam soil were lost with applications of calcium arsenate + organic matter, arsenic trioxide + organic matter, and arsenic trioxide alone, respectively. Incubation time was eight weeks. Steevens, et al (13) could not account for all of the sodium arsenite applied to a Plainfield sand (0-83 cm profile). Loss by wind erosion was suggested but methylarsines may have geen generated.

Hiltbold, et al (26) were unable to recover 67, 57 and 39% of the arsenic applied as MSMA to Hartsells fine sandy loam, Decatur silt loam and Dothan loamy sand soils, respectively, after six years of application.

The loss values calculated for Transfer 7 were in addition to the oxidation of methylarsines back to the original site of application. It can only be concluded that the net loss of methylarsines was greater than the value for this transfer; this pathway could be influenced by air movement, precipitation and the initial concentration of methylarsines in air.

Transfer 9: Arsenic Input with Irrigation. It has been established that background levels of arsenic in freshwater do exist (35) (36) (37) (74). This input would be significant in the overall arsenic cycle, while being more important for certain areas and crops.

Salman, et al (54) estimated that four irrigations with water containing 0.05 ppm As would contribute 0.002 lbs of elemental As or 9.0 gms As/ha/yr. This calculation was exaggerated to present conditions that most favor soil build up; the average As level in water is well below 0.05 ppm.

Transfer 10: Movement of Arsenicals from Treated Soil to Water. The two mechanisms considered for this transfer were 1) leaching of arsenicals through the soil profile to ground water, and 2) wind and water erosion of treated soil particles directly to water sources. The arsenic fixing capacity of soils has been previously discussed in Sections A-I and A-II. It was concluded that the most influential factors were reactive levels of Al and Fe in soil but that phosphorus additions, soil type, type of clay, and pH may affect Al and Fe availability.

1) Leaching. Jones and Hatch (2) surveyed old orchard soils in Oregon that were unproductive due to high levels of arsenic applications as insecticides. A clay adobe soil contained 441 ppm As in the surface 20.3 cm but only 19 ppm As in the next 20.3 cm; similarly, a silt loam soil was found with 61 and 7.6 ppm As at each respective depth. Benson (7) found that arsenic in orchard soils treated with lead arsenate was not readily leached below the root zone with percolating water. Vandecaveye, et al (3) sampled old orchard soils to a depth of 121.9 cm and found that the amount of water soluble arsenic was concentrated in the surface 15.2 cm with very little arsenic contained in soil below

61.0 cm.

Arnott and Leaf (24) concluded that arsenic trioxide applications at rates ranging from 1,120-8,960 kg/ha to a Chenango silt loam soil did not increase arsenic residues in the soil profile below 35.6 cm. The amount of leachate used affected residues in the surface 25.4 cm only.

Steevens, et al (13) recovered arsenic that was leached to a depth of 38 cm after treatments of 90 and 180 kg As/ha/yr (as sodium arsenite) to a Plainfield sandy soil; an application of 720 kg As/ha caused leaching of arsenic to 68 cm. This distribution was observed three years after application and in an area with relatively high precipitation (Hancock, Wisconsin).

Research by Dickens and Hiltbold (31) revealed that DSMA was not leached below 15.2 cm in a Decatur clay loam soil and that approximately 52% of the DSMA applied was moved through a 22.9 cm column of Norfolk loamy sand soil. The amount of DSMA applied was equivalent to 112 kg/ha and leaching was achieved with 50.8 cm of water. Soil pH did not affect leaching in this study.

Ehman (79) reported the results of leaching experiments with about 150 cm of water, through columns (about 30 cm) of sandy and sandy loam soils, containing DSMA (31.4 kg/ha) and CA (16.8 kg/ha) applied to the soil surface. The leachate removed about 9% of each product from the sandy soil and about 6% from the sandy loam soil.

Studies by Johnson and Hiltbold (28) concluded that all of the arsenic applied with 16 applications of DSMA or MSMA over a four-year period was within the upper 30 cm of a Chesterfield sandy loam soil; the rate of application was 8.92 kg/ha/yr with either treatment. Increased application rates of 17.88 and 35.80 kg/ha/yr of either chemical caused about 25 and 75% of the arsenic applied, respectively, to move below 30 cm in depth. Soil samples at 90 cm indicated that leaching had not occurred to that depth. Approximately 580 cm of rainfall was measured during the study period.

MSMA applications at rates as large as 40 kg/ha/yr over a six-year period did not cause leaching of arsenic below 30 cm. In this study, Hiltbold, et al (26) concluded that movement of arsenic was least in a Decatur silt loam, greatest in a Hartsells fine sandy loam, and intermediate in a Dothan loamy sand soil.

Norris (48) studied the movement of MSMA and CA in Klicker silt loam, Astoria silty clay loam and Edds loam soils from forest floors in the Northwest. MSMA was not leached through any of these soils (7.62 cm soil columns) while 86.4 cm of water were required to move 84 and 50% of the CA applied in the Klicker and Edds soils, respectively. It was concluded that neither chemical presented contamination hazards to groundwater.

Total doses of MSMA at 40.2 kg/ha and CA at 50.4 kg/ha over a six-year period have not increased arsenic levels below the plow layer in a Hidalgo sandy clay loam soil (25).

These studies indicated that the possibility of contamination of groundwater with leaching of arsenicals is very insignificant. All of the conditions which favor leaching (high rates of arsenic applied, light soils, and heavy leaching pressures) were tested, but leaching below the 90 cm depth of soil was not observed; in most studies this depth was considerably less. For the agronomic ecosystem model, this part of Transfer 10 was considered negligible.

2) Erosion. Erosion of soils has been conservatively estimated at 2.78 tons per acre per year when corn, wheat and clover were grown in rotation (80). If the average soil arsenic level is assumed to be 10 ppm As, approximately 60.0 gms As/ha/yr are lost with this transfer.

Transfer 11: Arsenic Inputs with Commercial Fertilizer.

Schroeder and Balassa (38) noted that phosphate fertilizers contained as much as 4.50 and 4.75 ppm As. This is not unexpected because of the similarities and interactions between As and P (38) (43) (27). Assuming that a soil test for phosphorus is within average values, a rate of about 84 kg/ha phosphorus is recommended to maintain soil fertility (81). A transfer value of 0.4 gms As/ha/yr was calculated for the model.

Transfer 12: Arsenic Contamination from Non-Agricultural Sources. Inputs of arsenic from the burning of fossil fuels (43), industrial wastes (35), and municipal wastes (27) are known to occur, but the amounts contaminating the agricultural ecosystem are not known at this time. The transfer rate for this model is assumed to be zero.

E. Conclusions

The ubiquity of arsenic in the environment is evidence of redistribution processes that have been operating since early geological times. Recent work by Braman and Foreback (75) confirms the presence of a natural arsenic cycle, but long-term effects of man-made arsenic inputs on this cycle remain uncertain.

This model was proposed to give some perspective on the impact of organic arsenical herbicides in an agronomic ecosystem; however, many of the transfers involved would certainly be applicable to other ecosystems. The reduction of arsenicals to methylarsines, soil erosion and arsenic uptake were the primary redistribution mechanisms in this model.

Since the soil is the ultimate sink for arsenic, a budget for soil was arranged. Values presented are theoretical and reflect conditions which favor a buildup of arsenic in soil. Net input into the soil with the CA model was 9,139.3 gm As, net losses were 1,721.7-3,365.0 gm As/ha/yr, giving a theoretical buildup of 2.6-3.3 ppm As/ha/yr. In the MSMA model, input was 5,512.3 gm, losses were 1,107.7-5,512.3 gm As/ha/yr, or a 1.5-1.9 ppm As gain.

The 17-35% values used for methylarsine losses may be low. These values were taken from studies in which herbicides were applied directly to bare soil surfaces. In practice, efficient applications of organic arsenical herbicide will distribute most of the residues in or on weedy plant tissue, increasing the possibility of methylarsine losses directly from decomposing vegetation to air and bypassing soil in the transfer sequence. It has been reported that garlic odors were most noticeable when high rates of CA or MSMA were used, so the percent lost may increase with continued applications. It is suspected that an equilibrium will occur in soil at which methylarsine losses equal organic arsenic inputs, but this has not been clearly demonstrated yet.

From the data presented in this model, it was concluded that arsenic is mobile and nonaccumulative in the air, plant and water phases of the agronomic ecosystem. Arsenicals do accumulate in soil but redistribution mechanisms preclude hazardous accumulations at a given site.

F. Literature Cited

1. Dorman, C., Tucker, F.H., Coleman, R., J. Amer. Soc. Agron. (1939) 31 1020-28.
2. Jones, J.S., Hatch, M.B., Soil Sci. (1937) 44 37-63.
3. Vandecaveye, S.C., Horner, G.M., Keaton, C.M., Soil Sci. (1936) 42 203-213.
4. Crafts, A.S., Rosenfels, R.S., Hilgardia (1939) 12(3) 177-200.
5. Ball, E.D., Titus, E.G., Greaves, J.E., J. Econ. Entom. (1910) 3 187-197.
6. Bishop, R.F., Chisholm, D., Can. J. Soil Sci. (1961) 42(1) 77-80.
7. Benson, N.R., Wash. State Hort. Assoc. Proc. (1968) 1E-5E.
8. Leaf, A.L., Smith, R.E., Jr., Weed Sci. (1960) 8 374-378.
9. Woolson, E.A., PhD thesis, Univ. of Maryland (1969).
10. Woolson, E.A., Axley, J.H., Kearney, P.C., Soil Sci. Soc. Amer. Proc. (1971) 35(6) 938-943.
11. Miles, J.R.W., J. Agri. Food Chem. (1968) 16(4) 620-622.
12. Williams, K.E., Whetstone, R.R., USDA Tech. Bull. 732 (1940) 20 pp.
13. Steevens, D.R., Walsh, L.M., Keeney, D.R., J. Environ. Quality (1972) 1(3) 301-303.
14. Steevens, D.R., Walsh, L.M., Keeney, D.R., Pest. Monitoring J. (1972) 6(2) 89-90.
15. Jones, J.S., Hatch, M.B., Soil Sci. (1945) 60 277-288.
16. Isensee, A.R., Shaw, W.C., Gentner, W.A., Swanson, C.R., Turner, B.C., Woolson, E.A., Weed Sci. (1973) 21(5) 409-412.
17. Woolson, E.A., Axley, J.H., Kearney, P.C., Soil Sci. Soc. Amer. Proc. (1973) 37(2) 254-259.
18. Small, H.G., McCants, C.B., Agron. J. (1962) 54 129-133.
19. Rosenfels, R.S., Crafts, A.S., Hilgardia (1939) 12(3) 203-229.

20. Hurd-Karrer, A.M., USDA Tech. Bull. 911 (1946) 31 pp.
21. Benson, N.R., Soil Sci. (1952) 74(3) 215-224.
22. Woolson, E.A., Weed Sci. (1973) 21(6) 524-527.
23. Jacobs, L.W., Syers, J.K., Keeney, D.R., Soil Sci. Soc. Amer. Proc. (1970) 34 750-754.
24. Arnott, J.T., Leaf, A.L., Weed Sci. (1967) 15(2) 121-124.
25. Sandberg, G.R., Allen, I.K., Dietz, E.A., Jr., Ansul Progress Report (1973) 82 pp.
26. Hiltbold, A.E., Hajek, B.F., Buchanan, G.A., Weed Sci. (1974) 22(3) 272-275.
27. The Ansul Co., Response to FR As and Pb Notice 36 FR 12709 (1971) 55 pp.
28. Johnson, L.R., Hiltbold, A.E., Soil Sci. Soc. Amer. Proc. (1969) 33 279-282.
29. Hamilton, K.C., Arle, H.F., Agri. in Ariz. (1968) 21(2) 16-17.
30. Schweizer, E.E., Weed Sci. (1967) 15(1) 72-76.
31. Dickens, R., Hiltbold, A.E., Weed Sci. (1967) 15(4) 299-304.
32. Frost, D.V., "The Arsenic Cycle in Nature," unpublished.
33. Bowen, H.J.M., "Trace Elements in Biochemistry," Academic Press, New York (1966).
34. Hertzog, E.S., Ind. Eng. Chem. (1935) 7(3) 163-165.
35. Durum, W.H., Hem, J.D., Heidel, S.G., Geo. Survey Circ. 643, 49 pp.
36. Kopp, J.F., Kroner, R.C., "Trace Metals in Waters of the United States," U.S. Dept. of Interior (1967).
37. Lis, S.A., Hopke, P.K., Environ. Letters (1973) 5(1) 45-51.
38. Schroeder, H.A., Balassa, J.J., J. Chron. Dis. (1966) 19 85-106.
39. Corneliussen, P.E., Pest. Monit. J. (1972) 5(4) 313-330.
40. LeBlanc, P.J., Jackson, A.L., Marine Pollution Bull. (1973) 4(6) 88-90.
41. Liebig, G.F., Jr., "Diagnostic Criteria for Plants and Soils," 13-23, Univ. of Calif. (1966).
42. Liebig, G.F., Jr., Bradford, G.R., Vanselow, L.A.P., Soil Sci. (1959) 88(6) 342-348.
43. Frost, D.V., Federation Proc. (1967) 26(1) 194-207.
44. Allaway, W.H., Adv. Agron. (1968) 20 235-274.
45. Isensee, A.R., Kearney, P.C., Woolson, E.A., Jones, G.E., Williams, V.P., Environ. Sci. Tech. (1973) 7(9) 841-845.
46. Macklin, C.E., Witkamp, M., MS thesis, Emory Univ., Atlanta (1973) 35 pp.
47. Greaves, R.D., MS thesis, Oregon State Univ. (1974) 94 pp.
48. Norris, L.A., USDA Forest Service report, Corvallis, Oregon (1974) 98 pp.
49. Malone, C.R., Soil Biol. Biochem. (1971) 3 127-131.
50. Malone, C.R., Ecology (1972) 53(3) 507-512.
51. Edwards, A.C., MS thesis, Auburn Univ. (1973) 90 pp.
52. Hunter, J.H., Young, A.L., USAF Tech. Rep. AFATL-TR-72-31 (1972) 25 pp.

53. Pate, B.D., USAF Tech. Rep. AFATL-TR-72-72 (1972) 13 pp.
54. Salman, H.A., Bartley, T.R., Summers, A.D., Bur. of Reclam. REC-ERC-72-37 (1972) 8 pp.
55. Duble, R.L., Holt, E.C., McBee, G.G., Weed Sci. (1968) 16(4) 421-424.
56. Holt, E.C., Faubian, J.L., Allen, W.W., McBee, G.G., Weed Sci. (1967) 15(1) 13-15.
57. Wills, G.D., Proc. So. Weed Sci. Soc. (1970) 23 334-337.
58. Keeley, P.E., Thullen, R.J., Weed Sci. (1970) 18(4) 437-439.
59. Keeley, P.E., Thullen, R.J., Weed Sci. (1971) 19(5) 601-606.
60. Duble, R.L., Holt, E.C., Weed Sci. (1970) 18(1) 174-179.
61. Cooley, A.W., MS thesis, Texas Tech. Univ. (1971) 63 pp.
62. Sckerl, M.M., Frans, R.E., Weed Sci. (1969) 17(4) 421-427.
63. Ehman, P.J., Proc. 19th Ann. Meeting SWC (1966) 540-541.
64. Duble, R.L., Holt, E.C., McBee, G.G., Agri. Food Chem. (1969) 17(6) 1247-1250.
65. Keeley, P.E., Thullen, R.J., Weed Sci. (1971) 19(3) 297-300.
66. Vallee, B.L., Ulmer, D.D., Wacker, W.E., AMA Arch. Ind. Health (1960) 21 132-151.
67. Thom, C., Raper, K.B., Science (1932) 76(1980) 548-550.
68. Challenger, F., Higginbottom, C., Ellis, L., J. Chem. Soc. (1933) Trans. 95-101.
69. Challenger, F., Higginbottom, C., Biochem. J. (1935) 29 1757-1778.
70. McBride, B.C., Wolfe, R.S., Advan. Chem. Ser. (1971) 105 11-22.
71. Smith, H.R., Cameron, E.J., Ind. Eng. Chem. (1933) 5 400-401.
72. Zussman, R.A., Vicher, E.E., Lyon, I., J. Bacteriol. (1961) 81 157.
73. Merrill, W., French, D.W., Proc. Minn. Acad. Sci. (1964) 31(2) 105-106.
74. McBride, B.C. Wolfe, R.S., Biochem. (1971) 10(23) 4312-4317.
75. Braman, R.S., Foreback, C.C., Science (1973) 182 1247-1249.
76. Cox, D.P., Alexander, M., Bull. Envir. Contam. Toxic. (1973) 9(2) 84-88.
77. Woolson, E.A., Kearney, P.C., Envir. Sci. Tech. (1973) 7(1) 47-50
78. Epps, E.A., Sturgis, M.B., Soil Sci. Soc. Amer. Proc. (1939) 4 215-218.
79. Ehman, P.J., Proc. 18th Ann. Meeting SWC (1965) 685-687.
80. Buckman, H.O., Brady, N.C., "The Nature and Properties of Soil," 6th ed., p. 230, The Macmillan Co., New York (1967).
81. Tisdale, S.L., Nelson, W.L., "Soil Fertility and Fertilizers," 2nd ed., The Macmillan Co., New York (1968).

10

Simulation of the Mobility of Arsenic Compounds in the Environment: A Southern Texas Case Study

JOHN E. STOLZENBERG

Institute for Environmental Studies and Department of Chemical Engineering, University of Wisconsin, Madison, Wisc. 53706

In studying the behavior and effects of an economic poison or a family of economic poisons, one is ultimately concerned with the behavior of the material in natural systems, that is, in the field. Unfortunately, due to the complexities of such behavior the researcher is often forced to perform either laboratory experiments or highly controlled field experiments in order to hold his experimental design and observations to manageable proportions. Common practice then calls for experts to extrapolate these experimental results to predict the expected behavior of the chemical for much larger and more realistic natural systems. When many interacting parameters are involved, the experts often either consciously or unconsciously simplify their evaluation to be able to arrive at a conclusion. In this paper, the development is reported of a computer simulation model designed to integrate diverse information on the mobility of a selected chemical in a regional environment. Since it is impossible to derive all the mass transport equations describing such a complex, multidimensional system, the model has been formulated probabilistically rather than deterministically. A flow path is constructed for an indivisible portion of the chemical. Then flow path construction is repeated many times, and the distribution is averaged in order to predict the representative behavior of the chemical. To cope with the large data requirements, often on unknown aspects of the material's behavior, the model can incorporate either objective or subjective quantifications of the processes influencing the chemical being simulated. Thus, the model can be viewed as a tool which augments experts' understanding of the behavior of specific compounds in a selected system.

In the next section an overview of the model will be presented. This will be followed by a discussion of the case study that the model is being applied to. This case study involves predicting the fate of disodium methanearsonate (DSMA) and its derivatives after it has been applied by cotton farmers in Southeastern Hidalgo County, Texas.

Overview of the Model

Basic Model Logic. The model is based on the simple observation that for a portion of the chemical to get to a specified position and form (the state of the chemical) within the regional environment, certain recognizable transitions in a specific sequence had to have happened to this portion. Furthermore, once in a given state, only a limited number of things can then happen. For example, for a pesticide to be adsorbed onto soil particles ten centimeters beneath the soil surface, first the sprayed chemical had to be carried into the soil, possibly by leaching. Once adsorbed, this portion of the chemical could only be influenced by a limited number of processes such as desorption or uptake by microorganisms. These observations lead to the basic algorithm, or recipe, for the model: follow a portion of the chemical through the regional environment by focusing on the competing processes that may affect the chemical for any given set of conditions. Considering the availability of information and the resolution of the model, "competing processes" are those processes which (a) cannot be further divided into two or more processes, and (b) represent the set of probable processes which can influence the chemical in a given state. In other words, by examining the competing processes for a given state, choosing one of them to operate on the chemical, and then performing the process within the model, a subsequent state can be identified. If these steps are repeated for each new state until appropriate termination is reached, a flow path for the chemical is constructed. As expressed in Table I, these operations form the basis for the model logic.

Table I. Basic Model Logic Governing Transitions Between States

1. Determine Realistic Environmental Conditions
2. Determine Possible Competing Processes
3. Select a Process
4. Perform Prescribed Process

The environmental conditions of interest are determined by the current state of the chemical. For example, if the flow path has followed the material to it being attached to dust particles floating through the air, the model would not have to retrieve information on soil pH. Instead, information such as wind speed and wind direction would be estimated based on historical records. The information on environmental conditions is then used in Step 3, "Select a Process."

The inherent problem with Step 2, "Determine Possible Competing Processes," is that a computer can't generate these processes. The list of all possible processes has to be structured so that for any state, the set of competing processes is immediately apparent, and once a process has been chosen (thus defining a new state) the new set of competing processes is apparent. The graph to structure the processes properly has been designed, and an example of it is presented below in the discussion of the case study. The non-numeric computer programming needed to construct the process graph is too lengthy to discuss here, but it is presented elsewhere (1).

"Select a Process" reflects the quantification of the decision rule which determines how to partition the chemical among the competing processes. Unfortunately, the understanding of the mobility of trace chemicals in natural systems is such that very few mechanisms which control the partition are known quantitatively. To circumvent this uncertainty, the model has been formulated probabilistically with the four steps in Table I applying to an indivisible portion of the chemical. This implies that at any given state, only one of the competing processes will act upon the chemical. To quantify this selection, each process must be described in terms of a probability function where the probability of the process occurring is a function of the properties of the chemical and environmental parameters influencing the process (as determined in Step 1). Then, the process to follow is chosen randomly. The likelihood of picking a process is weighted according to the size of the probability of that process occurring. This random selection is demonstrated in Figure 1 for a set of three competing processes. The probabilistic computational scheme using random numbers to choose outcomes just described constitutes a Monte Carlo simulation (2). The formulation of the probability functions describing the processes will be presented later in this paper.

The final step in constructing one transition in a flow path is to perform the prescribed process within the model. This is essentially a bookkeeping step. The time that the process takes must be accounted for and the new state of the chemical must be identified. Also, as mentioned earlier, whether or not the flow path should be terminated must be determined in this step. A path can be terminated for any one of the following four reasons: (1) the model predicts that the chemical is transported beyond the boundaries of the regional environment being modeled; (2) the end of the modeling period is reached; (3) the model predicts that the chemical is in a "sink," such as being tightly adsorbed onto clay particles; and (4) no following processes are specified for a given state, reflecting a lack of knowledge of what could happen next.

In order to program this logic for a digital computer, the representation of the regional system must be discretized into a three dimensional spatial grid. The grid cell boundaries

Step 1. Evaluate probability functions.

Probability (Spray Drift into Air) = f(chemical and environmental parameters) = 0.25

Probability (Deposit onto Soil) = f(chemical and environmental parameters) - 0.40

Probability (Deposit onto Plant Leaves) = f(chemical and environmental parameters) = 0.35

Step 2. Use a random number to pick a process.

Let the random number = 0.35

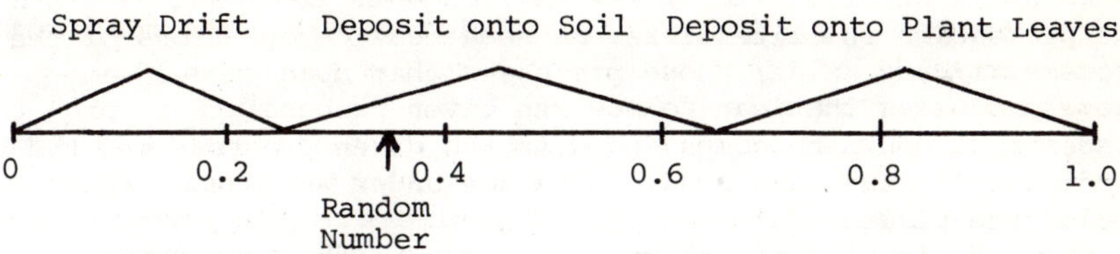

Therefore choose "Deposit onto Soil" for this Flow Path

Figure 1. Selecting a process

correspond to physical attributes of the system such as a water course bank or the edge of a soil series. Superimposed on this grid are functional pools and the various forms the chemical can be transformed to. The pools are where the transitions or changes to the chemical take place. An example of a pool for a herbicide is "Inside the Target Plants." Thus, the chemical can be transported among pools, or it can be transformed to a derivative within a pool.

Process Probability Functions. As described above, the model is a mechanistic simulation depending on both how the processes are ordered and quantified. In quantifying the processes, a balanced yet flexible treatment is called for so that neither transport nor transformation processes are biased. Three steps are followed in deriving a process probability function. To be included, each process must be identified as significant. Then, through a brief review, the process is broadly characterized. Finally, the process is quantified in a probability function.

Since it would be nearly impossible to incorporate every conceivable process that could affect the chemical(s) under study, only "significant" processes are included. A significant process is defined as a process that is proven or is strongly suspected by analogy with the behavior of other chemicals to be a major link in the flow path of the chemical in the case study situation. A link is major because either the amounts of material involved are relatively large or the resultant biological effects are important.

The second step is to review the known, pertinent characteristics of the process. This includes identifying the following: the major variables influencing the process, including properties of both the chemical and of the environment; important papers on the process in the literature; mathematical descriptions of the process, if any exist; those processes that must proceed and those processes that can follow the given process; those processes that could be competing with the given process; and the conditions in the case study situation under which the process would take place.

Quantification of the process can be either objective or subjective. If the information identified above is sufficient to describe the behavior of the chemical under a range of conditions for the regional environment of interest, then an objective probability distribution of the process can be formulated. But for most processes it is anticipated that little, if any, of the desired field information exists. In this case, experts' understanding of the process for the given situation will be quantified using techniques developed mainly by management scientists and psychologists (3). Here the concern is not so much how humans process information in reaching a decision as that subjective human estimates can be resorted to when objective probabilities cannot be formulated. The format followed in interviewing an expert is outlined in Table II and is based on the review by

Table II. Steps in an Interview Designed to Solicit an Expert's Subjective Probability of a Given Process Occurring

1. Initial Contact - The interviewer states his objectives and reviews the modeling technique and case study with emphasis on the role of probabilities.

2. Variable Selection - The interviewer and expert agree upon the major variables needed to describe the process and upon the processes competing with the one being quantified. Relevant research results, if available, are skimmed; ranges of the variables in the case study situation are identified. The expert ranks the variables in order of importance. Interactions among the independent variables are identified as are mass dependencies.

3. Solicit Probabilities - The interviewer solicits for specified sets of variables, the probabilities of the process occurring, or their equivalent, given the presence of the competing processes. The probabilities are associated with the same time interval for a complete set of competing processes. Experimental design techniques are used to insure adequate coverage of the ranges of the variables.

4. Consistency Checks - The interviewer spot-checks the expert's probabilities during Step 3 by providing feedback of the expert's previous results.

5. Follow-Up Questions - The expert is asked, for the conditions covered, how detailed (accuracy and collection periods) data on the variables should be and how applicable his process probability function is to other chemicals and other field conditions.

Huber (4). Once the experts' probability responses are obtained, they are formulated into a mathematical function using standard multiple regression analysis techniques. In addition to yielding a probabilistic description of the process, the information obtained from the experts will compliment that obtained in the characterization of the process.

To simplify the overall modeling effort, the key assumption is made that the probability of a process occurring is independent of the concentration of the chemical present in any given state being operated on. This assumption is based in part on the judgment that the occurrence of processes in the case study will be controlled more by environmental parameters such as soil moisture and pH than by the exact amount of the chemical present. Furthermore, the assumption seems warranted because the model will be more affected by whether or not a process is included in the simulation than by its exact mathematical description. Finally, at least for the case study, available historical records will be depended upon for the values of many of the environmental parameters. Since these records are of varying quality, a highly sophisticated simulation is not justified. Under this assumption, the probability of a process occurring equals the fraction of the chemical in the initial state that will be affected by the process. This assumption also implies that for any given state what happens to the indivisible portion of the chemical is independent of the past states that the chemical has been in. This is the "Markov Property," and it means that the computational scheme can also be considered a Markov Chain (1). Being a Markov Chain aids the interpretation of the model output. Now the proportion of flow paths ending in a given state after a specified time in the Monte Carlo simulation is equivalent to the fraction of the initially applied chemical in this state after the same time interval.

Arsenic Case Study

As introduced above, the general framework of the model is a way to conceptualize the behavior of a chemical. To proceed further with quantification, the model must be molded about a specific chemical or family of chemicals acting in a specific setting. The case study chosen for examination involves predicting the fate of the arsenical herbicide disodium methanearsonate (DSMA) as used by cotton farmers in Southeastern Hidalgo County, Texas. More precisely, the simulation objective is to predict where the DSMA applied during a representative cotton growing season will be and in what form after this cotton has been harvested. This region is about 50 kilometers (31 miles) up the Rio Grande River from Brownsville, Texas and is approximately 500 square kilometers (193 square miles) in size. Other characteristics of the case study region are summarized in Table III.

The processes felt to be significant have been identified, and now they are being characterized and quantified. An ordering

Table III. Summary of the Major Features of the Case Study Area, Southeastern Hidalgo County, Texas

Hydrology - Rio Grande River is the southern boundary; Rio Grande Floodway cuts through; North/South ditches are used to carry irrigation water; 51 to 76 centimeters (20 to 30 inches) of rain fall year (coming mainly in high intensity tropical storms).

Topography - The area is fairly flat, having an elevation change of 9 to 12 meters (30 to 40 feet) between the Northwest and Southeast corners.

Meteorology - Predominant winds are from the east; mean annual temperature is about 23°C (74°F).

Biota - Area is predominantly agriculture; main crops are cotton and sorghum, grown in similar proportions; grapefruit orchards and some truck farms are also present.

Soil - Types vary over approximately 20 series, including: "1) level, moderately and slowly permeable, loamy soils of flood plains and low terraces, 2) level, moderately and slowly permeable, loamy and clayey soils of uplands, 3) gentle sloping, moderately permeable, loamy soils of uplands, and 4) level, very slowly permeable, high shrink-swell clayey soils of low terraces and upland." (5); soil pH ranges from 6.6 to 8.4.

of these processes is under review. To understand these processes better, it is useful to first consider the pools where the behavior of the chemicals in the model is aggregated and the derivatives of the DSMA that are included in the simulation.

Pool Identification. As defined above, the pools are aggregates of functional locations of all the point phenomena affecting the chemicals. The use of pools helps in identifying processes and data requirements. By connecting a block diagram of the pools with arrows corresponding to processes, the flow chart of all considered paths is defined. (Such a chart would have to be at least two feet by three feet to be legible and comprehensive.) The seventeen pools defined for the case study are given in Table IV. They reflect a balance among manageability of the computational aspects of the model, realistic portrayal of the actual behavior of the arsenic compounds and limitations in experimental data.

Table IV. Pools Used in Representing the Case Study Environment

Spray Equipment

1. Spray Solution

Soil Medium

2. Ground Surface
3. Bulk Soil Water
4. Soil Particle Surface
5. Groundwater

Water Medium

6. Open Water Surface
7. Bulk Channel and Lake Water
8. Channel and Lake Bottom
9. Aquatic Food Web

Air Medium

10. Air

Terrestrial Biota

11. Target Plant Leaf Surface
12. Non-Target Plant Leaf Surface
13. Inside Target Plant
14. Inside Non-Target Plant
15. Harvested Crop
16. Terrestrial Food Web

Man-Made Structures

17. Cities

The starting point for all flow paths is the Spray Solution pool. This represents all of the DSMA applied to fields by spraying and related operations such as spillage onto fields. Arsenic compounds from other sources such as that transported by air into the region or herbicides sprayed on right-of-ways are not considered in the present version of the model. From the Spray Solution pool, the DSMA goes into one of the four surface pools or into the air.

The surface pools are quite important in the model because they represent interfacial areas where critical partitioning in the flow paths of the arsenic materials takes place. The DSMA is deposited onto the ground and plant leaves by the sprayer. The non-target plant is mainly cotton while weeds such as Johnson grass are the target plants. Spray drift carries the herbicide to either nearby water surfaces or into the air long enough for the DSMA to be converted to another form. Data limitations necessitate the aggregation of the behavior of the materials in the air into one pool.

Bulk Soil Water represents water in the soil that is not tightly bound, that is, moisture content above the wilting point. Here, vertical transport in the water column takes place. When irrigation is sufficient in the case study region to create a seasonal water table and the possibility of horizontal water movement, this behavior is represented in the Groundwater pool. The Soil Particle Surface pool is a combination of the microorganisms in the soil, of the surfaces of all the individual soil particles, and of the tightly bound interstitial soil water. These are lumped together because of the difficulties in experimentally separating them. Within this pool, gross phenomena such as adsorption and microbial oxidation take place. The space in the expanding lattices of certain clay minerals can be an important pool for some material. But, since the arsenic ions considered in the model are anionic in nature, this pool is not included.

In addition to their surfaces, the insides of both target and non-target plants are treated as pools. Here the concern is more with what arsenic goes into the plants and what comes out rather than dwelling on the biochemical reactions taking place within the plants. Harvested Crop is an important pool because it represents a pathway for the arsenic chemicals to humans. Cotton fiber and cotton seeds are parts of this pool. The final Terrestrial Biota pool is the conglomerate, Terrestrial Food Web (or Chain). This pool represents the uptake of arsenic compounds by animals and includes transfer of the chemicals by predator relations up a food chain. It is included even though there is little documentation of the arsenic herbicides so functioning.

The Cities pool is another route of exposure to man. The DSMA or its derivative could conceivably reach a city via air transport processes. The case study region has a string of small cities in it, which due to their location relative to the prevailing winds, could be contaminated by the arsenic compounds. If a flow path does indeed lead to a city, the model will not predict what happens to the chemical once it gets to the city.

The pools representing the Water Medium in addition to the water surface are straightforward. Bulk Channel and Lake Water include, besides rivers and lakes, irrigation ditches and floodways. The bottom or sediments of these waterways constitute another pool. Finally, the biological uptake is lumped into an Aquatic Food Web pool.

Forms of Arsenic. To completely understand the behavior of DSMA, it is necessary for the model to incorporate transformations of DSMA to other forms of arsenic as well as the transport of DSMA within the case study region. These transformations are assumed to occur entirely within pools. Transformations may also occur among the derivatives of DSMA, implying that the model is capable of simulating cycles among arsenic compounds. The model is programmed to consider the following forms in addition to DSMA: monosodium methanearsonate (MSMA), the salts of dimethylarsinic acid (cacodylic acid or CA), the arsines, and the salts of inorganic arsenates. The salts are used because of the alkaline nature of the case study area soil. Inorganic arsenites are not included.

MSMA is included in the model due to a combination of two reasons. First, measured by acute toxicity MSMA is less toxic than DSMA (6), implying differences in behavior between the two. Secondly, MSMA is more than likely being formed in the case region's soil. Since the soil pH ranges from 6.6 to 8.4 while the dissociation constant, pK_2, between MSMA and DSMA has been reported to be 7.82 (7), 8.24 (8), and 8.7 (9), appreciable proportions of DSMA could be converted to MSMA, depending mainly on the specific soil pH.

Evidence continues to mount that under proper conditions, various forms of arsenic can be biologically reduced to either arsine, methylarsine or dimethylarsine (10-12). Due to the analytical difficulties in distinguishing among these forms of arsines, all three will be considered together in the model.

Based on the author's limited review, there are no reports in the literature of DSMA or MSMA being methylated to cacodylic acid in field conditions. However, Wood proposed that this reaction does occur as part of the biological cycle for arsenic (13). A recent study analyzing environmental samples for methylated forms of arsenic concluded, "Dimethylarsinic acid is a major and ubiquitous form of arsenic in the environment," (14). In early lab studies, Dehn and Wilcox showed that dimethyl arsines are slowly oxidized to cacodylic acid (15). Finally, future versions of the model may also use cacodylic acid as a starting material. Therefore, for these reasons, it is felt that cacodylic acid is potentially an important form of arsenic in the case study and should be included in the modeling effort.

Under aerobic conditions, DSMA (10), MSMA (16) and Cacodylic Acid (11) have been shown to ultimately oxidize and demethylate to inorganic arsenates. Thus, it is necessary to include the arsenates as a significant arsenic form in the model.

One form of arsenic not included is inorganic arsenite. At the level of detail of the proposed model, arsenite appears to be an insignificant form of arsenic. Quastel and Scholefield found that arsenite is readily oxidized biologically to arsenate (17). Furthermore, Von Endt, et. al., detected only MSMA and arsenate and no arsenite in their work on the degradation of MSMA (16).

Proposed Significant Processes. The processes discussed in this section are matched to the case study conditions. In other settings, additional processes may be called for in describing the behavior of arsenic compounds. Some of the processes are included for the sake of completeness and will probably have to be compromised and aggregated with other processes due to data limitations. Yet it is useful to start with a fairly complete list of processes because it forces one to examine what he really knows about the total behavior of a family of chemicals. The literature citations are not exhaustive, but they are felt in most cases to be sufficient to identify a process as significant. When possible, the citation refers to a process affecting one of the five forms of arsenic included in the simulation. Transport processes imply transitions between two pools or cycling within one pool for one form of arsenic. For example, Spray Drift is between the Spray Solution pool and the Air pool, while Leaching can occur entirely within a pool. Transformation processes refer both to changes in physical states and to alteration to another arsenic form. Physical Adsorption illustrates the former; Biological Reduction the latter. Table V, which contains a summary of the proposed significant processes, serves as an outline to the introduction to the model processes which follows.

The Introductory Transport processes are the routes by which the DSMA enters the regional environment from the spray equipment. Recommended techniques and rates of applying DSMA are available from manufacturers (18). Depending upon the weather conditions at the time of spraying, drift of the liquid spray drops can occur (19, 20).

Changes in the physical states of a given arsenic species occur by physical transformations. Physical Adsorption and Desorption involving Van der Waals forces are one such process. The factors influencing adsorption have been reviewed by Bailey and White (21). Various studies, where the adsorption is correlated to the clay contents and thus the soil surface area, indicate that this process does influence arsenic compounds (10, 22). Desorption can often be inferred from leaching studies. Apparently only the arsines undergo Vaporization (23). What happens in field soils once the arsines are produced has not been reported even though Vaporization is potentially a critical link in cycling arsenic out of a fixed location in the soil. Work on measuring the Vaporization of other pesticides should be of help in the modeling effort (24-27). The Precipitation of insoluble arsenic compounds after Ion Exchange is the mechanism for another type of sorption phenomena, chemisorption. Pietsch has reported at what

Table V. Proposed Significant Processes

Introductory Transport Processes

- Deposit onto Soil and Plants
- Spray Drift into Air or onto Water

Physical Transform Processes

- Physical Adsorption
- Physical Desorption
- Vaporization
- Precipitation

Chemical Transform Processes

- Ion Exchange (Complexation)
- Dissociation
- Oxidation

Microbial Transform Processes

- Reduction (and Methylation)
- Oxidation (and Demethylation)

Plant Transport Processes

- Absorption into Leaves
- Root Uptake
- Plant Decay

Food Web Transport Processes

- Aquatic Organism Uptake
- Terrestrial Organism Uptake

Human Intervention Transport Processes

- Harvest Crop
- Plow in Plants
- Plow in and up Soil
- Pump Water from Fields

Plant Transform Process

- Metabolism by Plants

Water-Soil Transport Processes

- Leach into Soil and off Plants
- Water Erosion

Water Transport Processes

- Bulk Flow Transport
- Mixing by Turbulence
- Sedimentation
- Scour Pick-Up

Air Transport Processes

- Particle Transport by Wind and Turbulence
- Sedimentation
- Precipitation Scavenging
- Impaction
- Wind Erosion

pH's various metal ions will form a precipitate with methylarsonic acid and dimethylarsinic acid (28), and Chukhlantsev has reported the solubility products for a series of arsenates (29, 30).

In chemical transformations, the structure of the compound is altered. The arsenates (31, 32), MSMA (33), DSMA (10), and cacodylic acid (11) all have been reported to exchange or complex with iron, aluminum or calcium ions and oxides. With the basic pH's and high calcium levels in the case study soils, chemisorption with calcium complexes becomes one of the more important processes in the simulation. The Dissociation of methanearsonic acid between its monosodium and disodium salts was introduced in the discussion of the arsenic forms included in the model. At this time, the resolution of the model does not warrant distinguishing between the mono and disodium salts of arsenate. The equilibrium constant of cacodylic acid is low enough, 6.27 (8), that essentially most of the cacodylate will occur as a salt. Chemical Oxidation is apparently a relatively rapid decomposition mechanism of the arsines (34). On the other hand, cacodylic acid has been reported to be very resistant to Chemical Oxidation (14).

Wood recently pointed out the importance of microbial conversions of metals in the environment (13). He also proposed a biological cycle among arsenate, arsinite, methylarsonic acid, dimethylarsinic acid and the arsines based in part on the work of McBride and Wolfe (12). Various other workers have demonstrated the importance of Microbial Oxidation or Reduction with appropriate Demethylation or Methylation for DSMA (10, 22), MSMA (16), and cacodylic acid (11). The process of the microbes taking up a chemical is lumped in the model with the appropriate transformation that follows.

The Plant Transport processes describe mechanisms for moving the arsenic compounds either into or out of the two pools, Inside Target Plants and Inside Non-Target Plants. Even though the effects of the Absorption of sprayed arsenic herbicides into plant leaves has been studied (7, 35-37), few reports were found on the actual Absorption process. Since DSMA is sprayed with a surfactant added to the spray solution, Foy and Smith's review on the role of surfactants will be helpful (38) as will Ashton and Crafts introduction to the Absorption of arsenicals (39). Root Uptake of residual arsenicals is important because of the implications for understanding the ultimate loading, if any exist, of treated areas (31, 33, 40, 41). Plant Decay is the process whereby arsenic within a plant is carried to the soil by the dying and subsequent decaying of the plant.

Translocation within plants is lumped together with Metabolism by Plants in the model. At least in some species DSMA is not readily broken down (42), while it is strongly suspected that organic arsenicals are reduced to arsines within some plants (34). If plants do indeed produce arsines, this would provide an additional route for mobilizing arsenicals out of soil and plant pools.

The Water-Soil Transport processes occur when sufficient water interacts with plant surfaces or the soil in carrying an arsenic compound to or through a soil medium pool. Even though the arsenicals are usually strongly bound by chemisorption, the short distances Leaching carries the material can be important in introducing it to new reactive conditions (10, 22, 33). In spite of the case study region being fairly flat, the high intensity rain storms in the area can remove surface soil by Water Erosion. Therefore, any arsenical sorbed to this soil will also be transported. Work on water erosion and the removal of other materials will be useful in characterizing Water Erosion for this model (43, 44).

If an arsenic compound reaches a watercourse or lake, the Water Transport processes describe the movement of the material with the water and to and from the sediments. As a first approximation, the arsenical is assumed to be bound to a particle in the water so a standard treatment of inland hydrology should be sufficient to characterize these processes (45). Mixing by Turbulence refers to mixing within the water column while Sedimentation is a setting process carrying particles by gravity to the waterway bottom. Scour Pick-up is the removal of bottom particles by the bed load of sediments moving along the channel bottom.

In addition to Spray Drift, arsenicals can enter the air pool by Wind Erosion or possibly by Vaporization of arsines. If the latter process does take place as the evidence on biological reduction of arsenic compounds indicates, then Air Transport processes will be especially important in redistributing arsenic in the environment. No work on the air transport of arsenicals has been found in the literature, but a review of the concepts with respect to pesticides in general is available (46). Work by the United States Atomic Energy Commission also provides a firm introduction to these processes (47, 48). As with water, Sedimentation is the fall out of particles due to gravity. Precipitation Scavenging describes washing particles from the air both in and below clouds by rain (49), and it constitutes an important way of returning material from the lower atmosphere. Impaction is a deposition mechanism (50). It is of concern in the model as a contaminating mechanism of cities. Apparently a small amount of Wind Erosion does occur in Southeast Hidalgo County as evidenced by the appearance of small dust clouds even after crops have started to emerge from the soil (9). Wind Erosion can be characterized using factors in a "Universal Equation for Measuring Wind Erosion" (51).

The Food Web Transport processes describe both the uptake from the environment by organisms and the consumption of a prey containing an arsenical in its tissues by a predator. Here the concern is more with the organisms physically transporting the chemical to another location, leaving the interpretation of effects to biologists. In a model aquatic ecosystem, cacodylic acid was found to not magnify to any great degree through food chain feeding relations (52). In another setting, certain species of earthworms were found to play an important role in the vertical redistribution of cadium from the litter in the soil profile (53). The influence of soil insects and earthworms in the case study is unknown, but if present it would probably be important.

The final set of processes considered in the model describe transport mechanisms immediately resulting from the intervention of man. If residues of the arsenicals are present in the cotton fiber or seed when the cotton is harvested, then this suggests an important pathway of the material to man. By law, the cotton farmers have to plow in the cotton plant remaining in the field after harvesting by September 1st each year. Thus, any arsenic in the plant or on or near the soil surface will be bodily moved to 23 to 38 centimeters underground. This process potentially exposes the arsenic plowed in (as well as that plowed up) to a new set of reactive conditions. Such action more than likely greatly affects the long term behavior of arsenic residuals in the soil. During intense tropical storms, parts of the case study region are so flat that the rain water doesn't readily drain. To protect their crops from the effects of ponding, the local farmers often pump their fields dry, placing this water in irrigation ditches. If arsenicals are sorbed to soil particles near the soil surface of such flooded areas, it is conceivable that pumping could transport the chemicals to an aquatic system.

Process Graph Example. To demonstrate how the processes, pools and forms discussed above will be structured in the case study simulation, the process graph showing all the combinations of possible flow paths for DSMA is presented in this section. This graph indicates a transformation from DSMA to one of the four other forms of arsenic considered in the model, but it doesn't follow the flow path for the new form. When all forms of arsenic are included in the graph, it will be considerably longer. Also, the process graph for DSMA as given in Table VI will more than likely be modified as experts are consulted. The number for each entry is a unique identifier used in the computer programming of the graph. Next is a verbal description of the process. Following the slash is the form of the indivisible portion of arsenic present after the process has occurred. DSMA, MSMA, and CA are the standard abbreviations while ASO_4 stands for arsenate.

Since this ordering of processes forces one to be quite specific about the behavior of DSMA including unknown aspects, many assumptions are necessary. For example, it is assumed that microbes only take up DSMA from solution and that plowing only affects bound materials. Adsorption mechanisms are lumped together into one process. What happens when terrestrial or aquatic organisms take up DSMA is not specified, reflecting a lack of knowledge about these processes.

Figure 2 is presented to help the reader interpret Table VI. The process graph only gives processes (arrows in Figure 2) as the states between processes (boxes in Figure 2) are implied. The process graph can be viewed as an outline. Those processes with the same uninterrupted indentation and in a given sequence of process numbers are competing process. (Processes numbered 06 through 09 in Figure 2 are competing processes.) As one moves down and in one indentation from any given process, the subsequent set of competing processes is given. (Assuming 01 Deposit onto Soil is the first process chosen, then the set of competing processes following process 01 are processes 06 to 09.) The process described as "Sit" in the graph is added for bookkeeping purposes. When Sit is chosen as a process to follow, this implies that the chemical remains in its present state for a prescribed amount of time. Processes listed in parenthesis indicate the given flow path loops to another part of the process graph. (The set of competing processes that can occur after process 09 are processes 06 to 09). In this way, possible repeating of processes for more than one time step, feedback loops and cycling can easily be handled by the model. If no set of competing processes is listed beneath a given process, then this process is the terminal process for that particular flow path.

To illustrate how the model would use this process ordering a conceivable flow path for an indivisible portion of DSMA, assuming appropriate environmental conditions, is: (Spray Solution) 01 Deposit onto Soil (Ground Surface) 07 Adsorption (Ground Surface) 10 Water Erosion (Open Water Surface) 40 Mix into Bulk Channel (Bulk Channel Water) 47 Bulk Flow Transport (Bulk Channel Water) 46 Sedimentation (Channel Bottom) 52 Sit on Channel Bottom for one time step (Channel Bottom) 50 Microbial Oxidation (Channel Bottom). The processes are listed in the predicted order of occurrence. The pool that the chemical would be in after a process has occurred follows it in parenthesis. Eight random selections of processes among sets of competing processes using the Monte Carlo algorithm would have been used to create this flow path. It terminates because no processes are specified after 50 Microbial Oxidation.

By structuring the processes affecting the arsenic compounds in the case study environment in such a fashion, one explicitly defines all of the possible flow paths which the chemicals could follow. The quantitative simulation which follows the construction of the process graph takes these possible flow paths and gives the probable flow paths for any given set of environmental conditions. Furthermore, the process graph forces one to view each process in proper perspective, nestled between the process that must come before it and those that can follow it.

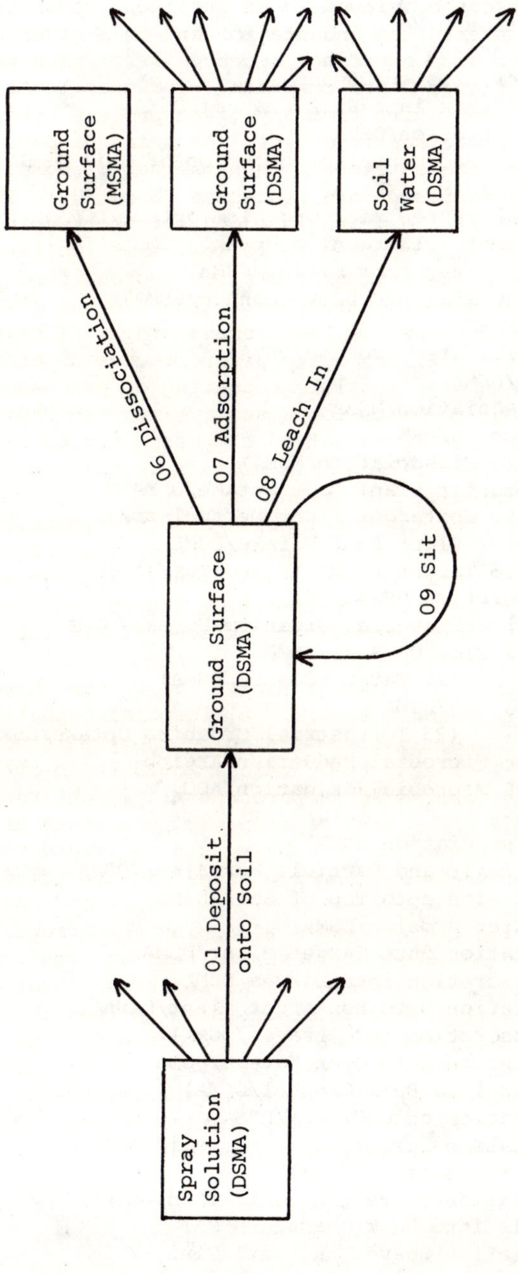

Figure 2. Elaboration of the beginning of the process graph given in Table VI

Table VI. Example Process Graph for DSMA

01 Deposit onto Soil/DSMA
- 06 Dissociation/MSMA
- 07 Adsorption/DSMA
  - 10 Water Erosion/DSMA
    - (40 Mix into Bulk Channel/DSMA)
  - 11 Wind Erosion/DSMA
    - (26 Sedimentation onto Top of Soil/DSMA)
  - 12 Plow in Soil/DSMA
    - (21 Terrestrial Organism Uptake/DSMA)
  - 13 Terrestrial Organism Uptake/DSMA
  - 14 Pump Water from Fields/DSMA
    - (40 Mix into Bulk Channel/DSMA)
  - 15 Sit/DSMA
    - (10 Water Erosion/DSMA)
- 08 Leach In/DSMA
  - 16 Dissociation/MSMA
  - 17 Leach In/DSMA
    - (16 Dissociation/MSMA)
  - 18 Nontarget Plant Root Uptake/DSMA
    - (62 Nontarget Plant Metabolism/?)
  - 19 Target Plant Root Uptake/DSMA
    - (56 Target Plant Metabolism/?)
  - 20 Adsorption/DSMA
    - 21 Terrestrial Organism Uptake/DSMA
    - 22 Plow Up Soil/DSMA
      - (10 Water Erosion/DSMA)
    - 23 Sit/DSMA
      - (23 Terrestrial Organism Uptake/DSMA)
    - 24 Microbial Reduction/Arsine
    - 25 Microbial Oxidation/ASO_4
- 09 Sit/DSMA
  - (06 Dissociation/MSMA)

02 Spray Drift in Air and Particle Formation/DSMA
- 26 Sedimentation onto Top of Soil/DSMA
  - (10 Water Erosion/DSMA)
- 27 Sedimentation onto Target Plant/DSMA
  - (53 Absorption into Leaves/DSMA)
- 28 Sedimentation onto Nontarget Plant/DSMA
  - (59 Absorption into Leaves/DSMA)
- 29 Sedimentation onto Open Waters/DSMA
  - (40 Mix into Bulk Channel/DSMA)
- 30 Sedimentation onto Cities/DSMA
- 31 Precipitation Scavenging onto Soil/DSMA
  - (10 Water Erosion/DSMA)
- 32 Precipitation Scavenging onto Open Water/DSMA
  - (40 Mix into Bulk Channel/DSMA)
- 33 Precipitation Scavenging onto Cities/DSMA

```
    34 Impaction onto Cities/DSMA
    35 Wind Transport/DSMA
        (26 Sedimentation onto Top of Soil/DSMA)
03 Spray Drift onto Nearby Open Water/DSMA
    36 Dissociation/MSMA
    37 Adsorption/DSMA
        40 Mix into Bulk Channel/DSMA
            (46 Sedimentation/DSMA)
        41 Flow Transport at Surface/DSMA
            (40 Mix into Bulk Channel/DSMA)
    38 Mix into Bulk Channel/DSMA
        42 Dissociation/MSMA
        43 Bulk Flow Transport/DSMA
            (42 Dissociation/MSMA)
        44 Adsorption/DSMA
            46 Sedimentation/DSMA
                49 Microbial Reduction/Arsines
                50 Microbial Oxidation/ASO4
                51 Scour Pick-Up/DSMA
                    (46 Sedimentation/DSMA)
                52 Sit/DSMA
                    (49 Microbial Reduction/Arsines)
            47 Bulk Flow Transport/DSMA
                (46 Sedimentation/DSMA)
            48 Aquatic Organism Uptake/DSMA
        45 Aquatic Organism Uptake/DSMA
04 Deposit onto Target Plant/DSMA
    53 Absorption into Leaves/DSMA
        56 Target Plant Metabolism/?
        57 Target Plant Decay/DSMA
            (10 Water Erosion/DSMA)
        58 Plow in Target Plant/DSMA
            (21 Terrestrial Organism Uptake/DSMA)
    54 Leach Off Target Plant/DSMA
        (06 Dissociation/MSMA)
    55 Sit/DSMA
        (53 Absorption into Leaves/DSMA)
05 Deposit onto Nontarget Plant/DSMA
    59 Absorption into Leaves/DSMA
        62 Nontarget Plant Metabolism/?
        63 Nontarget Plant Decay/DSMA
            (10 Water Erosion/DSMA)
        64 Plow in Nontarget Plant/DSMA
            (21 Terrestrial Organism Uptake/DSMA)
        65 Harvest Crop/DSMA
    60 Leach Off Nontarget Plant/DSMA
        (06 Dissociation/MSMA)
    61 Sit/DSMA
        (59 Absorption into Leaves/DSMA)
```

Interim Assessment of the Model

The modeling effort to date already has proven to be a systematic way of determining the information required to adequately quantify the regional behavior of one family of chemicals. Obviously, information is lacking to adequately describe many of the processes identified above. At this time, the following three areas especially need to be investigated to complete even a basic understanding of the total behavior of the arsenicals in the selected system:

1. The conditions of arsine production in plants and microorganisms and, once formed, the behavior of the arsines must be further clarified.

2. The behavior of arsenicals in the atmosphere is unknown though potentially quite important.

3. The role of human intervention such as plowing on "natural" processes is in need of study.

Once the model is built and tested, a sensitivity analysis of selected processes and environmental parameters should provide further insight into the behavior of the arsenicals. Other uses of the model include indicating sampling sites for monitoring programs and with proper interpretation helping to predict the likely biological effects resulting from applying the herbicides in a specific manner.

Most important though, the model is felt to be a plausible approach to treating the complex problem of describing the mobility of selected trace chemicals. The probabilistic treatment allows formulation of the simulation, and the use of subjective probabilities circumvents many of the data limitations. In this way, the model hopefully will become a tool to compliment the work of specialists in analyzing and predicting the mobility of arsenic compounds in the environment.

Acknowledgment

The author is grateful for the thoughtful assistance of T. W. Chapman, G. P. Huber, T. G. Kurtz, D. R. Keeney, E. F. Joeres and O. L. Loucks of the University of Wisconsin - Madison, and L. O. Moore, R. W. Olson, G. Sandberg, and P. J. Ehman of the Ansul Company, Marinette, Wisconsin.

This research was supported by funds from the National Science Foundation's Research Applied to National Needs program as administered by the Institute for Environmental Studies at the University of Wisconsin. The assistance of the Center for Resource Policy Studies and Programs and N. Erickson's help in preparing the manuscript is also appreciated.

Literature Cited

1. Stolzenberg, J.E., "Computer Aided Synthesis of the Environmental Pathways for Disodium Methanearsonate," Preliminary Research Proposal Submitted in Partial Fulfillment of the Degree of Doctor of Philosophy (Environmental Studies) at the University of Wisconsin-Madison, February, 1974. (Available from the author)
2. Naylor, T.H., et. al., "Computer Simulation Techniques," pp. 1-4, 333-335, John Wiley and Sons, New York, 1966.
3. Slovic, P. and S. Lichtenstein, Organizational Behavior and Human Performance (1971) 6, 649-744.
4. Huber, G.P., Decision Sciences (1974) 5, 430-458.
5. Soil Conservation Service, U. S. Department of Agriculture, General Soil Map #4-R-32792, Temple, Texas, 1973.
6. Christensen, H.E. (ed.), "Toxic Substances Annual List - 1971," 512 pp., National Institute for Occupational Safety and Health, U. S. Department of Health, Education, and Welfare, Rockville, Maryland, 1971. (Available from the Superintendent of Documents, #1716-0004)
7. Dub, M. (ed.), "Organometallic Compounds - Methods of Synthesis, Physical Constants and Chemical Reactions; Volume III, Compounds of Arsenic, Antimony and Bismuth," 2nd Edition, Springer-Verlog, New York, 1968.
8. Doak, G.O. and C. D. Freedman, "Organometallic Compounds of Arsenic, Antimony, and Bismuth," pp. 27, 31, Wiley-Interscience, New York, 1970.
9. Moore, L.O., Personal Communication, June 6, 1974.
10. Akins, M.B., M.S. Thesis, 107 pp., University of Tennessee, Knoxville, 1973.
11. Woolson, E.A. and P.C. Kearney, Environmental Science and Technology (1973) 7, 47-50.
12. McBride, B.C. and R.S. Wolfe, Biochemistry (1971) 10, 4312-4317.
13. Wood, J.M., Science (1974) 183, 1040-1052.
14. Braman, R.S. and C.C. Foreback, Science (1973) 182, 1247-1249.
15. Dehn, W.M. and B.B. Wilcox, American Chemical Journal (1906) 35, 1-41.
16. Von Endt, D.W., P.C. Kearney, and D.D. Kaufman, Agricultural and Food Chemistry (1968) 16, 17-20.
17. Quastel, J.H. and P.G. Scholefield, Soil Science (1953) 75, 279-285.
18. The Ansul Company, Specimen Label for Ansar Brand DSMA Liquid, 2 pp., Marinette, Wisconsin.
19. Seymour, K.G., Advances in Chemistry Series (1969) 86, 135-154.
20. Maybank, J. and K. Yoshida, in J.T. Bergsteinsson and W. Baier (eds.), "Meteorological Aspects of Pollution in Relation to Agricultural Pesticides," pp. 13-24, Canada Committee

on Agricultural Meteorology, Canada Department of Agriculture; Ottawa, 1971.
21. Bailey, G.W. and J.L. White, Residue Reviews (1970) 32, 30-80.
22. Dickens, R. and A.E. Hiltbold, Weeds (1967) 15, 299-304.
23. Weed Science Society of America, "Herbicides Handbook," 2nd ed., pp. 23, 252, 261, W. F. Humphrey Press, Geneva, New York, 1970.
24. Parmele, L.H., E.R. Lemon and A.W. Taylor, Water, Air and Soil Pollution (1972) 1, 435-451.
25. Hartley, G.S., Advances in Chemistry Series (1969) 86, 115-134.
26. Hamaker, J.W. and H.O. Kerlinger, Advances in Chemistry Series (1969) 86, 39-54.
27. Spencer, W.F. in Pesticide Research Center, "Pesticides in the Soil: Ecology, Degradation and Movement," pp. 120-128, Michigan State University, East Lansing, 1970.
28. Pietsch, R., Midrochim Acta (1958) 2, 220-224.
29. Chukhlantsev, V.G., Chemical Abstracts (1957) 51, 7107a.
30. Chukhlantsev, V.G., Journal of Inorganic Chemistry, U.S.S.R. (1956) 1, 1975-1982.
31. Woolson, E.A., J.H. Axley and P.C. Kearney, Soil Science Society of America Proceedings (1973) 37, 254-259.
32. Jacobs, L.W., J.F. Syers and D.R. Keeney, Soil Science Society of America Proceedings (1970) 34, 750-754.
33. Johnson, L.R. and A.E. Hiltbold, Soil Science Society of America Proceedings (1969) 33, 279-282.
34. Sandberg, G., "The Fate of Organic Arsenicals in an Agronomic Ecosystem: A Proposed Cycle," Weed Science Society of America Meeting Paper, Las Vegas, February 12-14, 1974.
35. Millhollon, R.W., Weed Science (1969) 17, 370-373.
36. Wiese, A.F. and E.G. Hudspeth, Jr., "Effects of DSMA and MSMA on Cotton Yield and Arsenic Content of Cottonseed," Texas Agricultural Experimental Station, Bulletin MP-877, College Station, 1968.
37. Baker, R.S., et. al., Weed Science (1969) 17, 37-40.
38. Foy, C.L. and C.W. Smith, Advances in Chemistry Series (1969) 86, 55-69.
39. Ashton, F.M., A.S. Crafts, "Mode of Action of Herbicides," 504 pp., Wiley-Interscience, New York, 1973.
40. Schweizer, E.E., Weeds (1967) 15, 72-76.
41. Woolson, E.A., Weed Science (1973) 21, 524-527.
42. Duble, R.L., E.C. Holt and G.G. McBee, Weed Science (1968) 16, 421-424.
43. Haan, C.T., Transactions of the American Society of Agricultural Engineers (1971) 14, 445-449.
44. Wischmeier, W.H., Agricultural Engineering (1962) 43, 212-215.

45. Linsley, R.J., Jr., M.A. Kohler and J.L.H. Paulhus, "Hydrology for Engineers," 340 pp., McGraw Hill, New York, 1958.
46. Bergsteinsson, J.T. and W. Baier (eds.), "Meteorological Aspects of Pollution in Relation to Agricultural Pesticides" 70 pp., Canada Committee on Agricultural Meteorology, Canada Department of Agriculture, Ottawa, 1971.
47. Slade, D.H. (ed.), "Meteorology and Atomic Energy 1968," 445 pp., Atomic Energy Commission Office of Information Services, 1968. (Available as TID-24190 from the National Technical Information Service)
48. Reiter, E.R., "Atmospheric Transport Processes, Part 2 Chemical Tracers," 382 pp., Atomic Energy Commission Division of Technical Information, 1971. (Available as TID-25314 from the National Technical Information Service)
49. Engelmann, R.J. and W.G.N. Slinn (Coords.), "Precipitation Scavenging (1970)," 499 pp., Atomic Energy Commission Division of Technical Information, 1970. (Available as CONF-700601 from the National Technical Information Service)
50. Mukammal, E.I. and G.A. McKay in J.T. Bergsteinsson and W. Baier (eds.), "Meteorological Aspects of Pollution in Relation to Agricultural Pesticides," pp. 33-47, Canada Committee on Agricultural Meteorology, Canada Department of Agriculture, Ottawa, 1971.
51. Agricultural Research Service, "A Universal Equation for Measuring Wind Erosion - An Aid to Conservation Farming in the Great Plains, ARS 22-69," U. S. Department of Agriculture, 1961.
52. Isensee, A.R., et. al., Environmental Science and Technology (1973) 7, 841-845.
53. Fulkerson, W., W.D. Schults, and R.I. Van Hook (eds.), "Ecology and Analysis of Trace Contaminants Progress Report, January 1973 - September, 1973," p. 115, Oak Ridge National Laboratory Report ORNL-NSF-EATC-6, Oak Ridge, 1974.

INDEX